D1236451

Inferential Statistics for Geographers
An Introduction

G. B. Norcliffe's career as a geographer was fixed at the age of five when his primary school teacher presented him with an 1869 edition of Philip's World Atlas. For the next ten years he was puzzled by the absence of such countries as Prussia and the Austro-Hungarian Empire from other atlases. At Cambridge, Glen Norcliffe was much persuaded by Peter Haggett's viewpoint that geography had a scientific basis, and that statistical methods were invaluable in geographical analysis. At Toronto University, and then at Bristol University, he specialized in economic geography, combining this with a continuing interest in statistical techniques. In 1970 he joined the Geography Faculty at York University, where, for a number of years, his teaching has included an introductory course in statistics for geographers. In recent years he has become increasingly concerned with planning issues, both in Canada, and as a planning advisor to the Kenyan government.

Geography editor
Michael Chisholm

Professor of Geography
in the University of
Cambridge

Inferential Statistics for Geographers

G. B. Norcliffe

Associate Professor Department of Geography
University of York, Ontario

A Halsted Press Book

JOHN WILEY & SONS
New York

Published in the U.S.A.
by Halsted Press, a Division
of John Wiley & Sons, Inc.
New York.

First published 1977
© G. B. Norcliffe 1977

Printed in Great Britain

Library of Congress Cataloging in Publication Data

Norcliffe, G   B
  Inferential statistics for geographers.
Includes bibliographies and index.
  1. Statistics.   2. Geography—Statistical methods.

I. Title.
HA29.N778   1977   519′.02′491   77–9427
ISBN 0–470–99206–9

To Mary

Contents

List of Figures

Preface

This book is written for university and college level geography students taking their first course in statistics. In my experience many such students have little confidence in their statistical abilities. However, I am convinced that this lack of confidence is largely unwarranted since only a few students can excuse themselves on grounds of mathematical incompetence. All that is assumed in this text is the ability to perform basic arithmetic operations: knowing how to look up logs, sines and cosines also comes in useful at one or two points.

The main problem for beginners is unfamiliarity with statistical concepts. Accordingly, the first objective in this book is to present statistical concepts and methods in a straightforward fashion: no mathematical proofs are provided, although where appropriate I have tried to give the reader an intuitive notion of the mathematical bases of certain techniques. Each technique is demonstrated by an example drawn from the real world.

The second objective in this book is to give the reader a geographical perspective. For instance, mathematical statisticians teaching introductory courses tend to put little emphasis on the Poisson distribution, whereas geographers give it greater attention in view of its importance to the discipline. Similarly many topics in spatial analysis, including centrographic measures, tests of contiguity and directionality, map comparisons, ecological correlations etc., can properly be included in an introductory statistics course for geographers, but are rarely mentioned in courses taught outside geography departments.

The third objective is to present statistics in such a way that the reader is immediately able to apply them. This has been done by adopting a hypothesis-testing framework. Null and research hypotheses are stated and formally tested, using a standard procedure that includes examples which are designed to illustrate problems commonly encountered by geographers. The result is to give less attention to point and interval estimates — not because they are unimportant or inappropriate, but simply on grounds of pedagogic convenience.

The final goal is to make the reader aware of the difficulties encountered in using statistical techniques. The assumptions associated with each method are stated, and in some cases discussed at length. Hopefully, the outcome will be an appreciation of the strengths and limitations of the various methods.

Much of the material presented in this text has been taught in an introductory statistics course for geographers that I have offered for several years at York University, Ontario. I am grateful to the students taking this course for their reactions and comments: the result has undoubtedly been to make the contents of this book more palatable. My greater debt is to my teachers — in particular Peter Haggett, Emilio Casetti, Leslie Curry and David Harvey — and to my colleagues — Roly Tinline, Ian Evans, Bob Murdie, Andrew Cliff, Keith Bassett and Rod White — who have played an important part in shaping my conception of statistics. I would also like to thank Arnold Court for permitting me to use his map comparison test and for raising a number of interesting points in correspondence, Conrad Heidenreich for allowing me to use some unpublished data in Chapter 6, and Michael Chisholm who has been a patient and helpful editor. The text has undoubtedly become less turgid as a result of a thoughtful editing by my wife, Mary.

The completion of this book is in no small part due to my typists Alice Leung, Florence Davies and Geraldine McLetchie: typing equations, and frequently having to change from Greek to Roman golf balls must indeed be a trying task. Finally, Carol Randall must be complimented for converting my rough sketches into professional diagrams.

May 1976

G. B. NORCLIFFE
Nairobi

Part one

Elements of statistical inference

1   Introduction

When one brings common sense to bear upon a problem, a mixture of
experience and intuition is used. When one brings inferential statistics to
bear upon a problem, one goes through a similar process, substituting
data for experience, and statistical formulae for intuition. Hence,
although the newcomer to statistics often thinks that he or she is enter-
ing a strange new world, in practice statistical methods involve no more
than doing — in a much more formal and rigorous way — the things
that are done informally countless times each day.

The newcomer will encounter many new terms and concepts, some of
which are discussed in this introductory chapter since they logically
precede the statistical tests discussed later. These preliminaries relate to
two basic questions: Why does one use statistics? And how does one go
about selecting a statistical test?

Why use statistics?

It is generally agreed that the overriding objective of all the sciences, the
physical, the biological and the social, is to *explain* events, to answer
the questions 'why', 'how', 'when', and 'where'. The bald statement of a
fact is not in itself very useful if no explanation is involved. Admittedly,
there are geographers who do not see the discipline as a science, and who
therefore do not place emphasis upon explanation, but the majority of
geographers probably agree that geography is either a social science or
an environmental science, or both, and that the ultimate objective of
geography is to provide rational explanations of the spatial attributes
and interrelationships of phenomena.

How do geographers set about explaining things? This question will
evoke quite different responses depending upon the person to whom it
is posed, these differences revealing a fundamental philosophical
dichotomy within the discipline. On the one hand are people concerned
to explain unique events largely in terms of their historical antecedents;
on the other hand are people concerned to make general statements

explaining sets of events. Coining terms to describe these two approaches is not easy, but those used by Hartshorne (1959, p. 149) now have fairly common currency; the *idiographic* approach is concerned with the study of individual events, while the *nomothetic* approach is concerned with the search for statements that have some general applicability.

As early as 1953 Schaefer criticized the exceptionalist tradition in geography, but it was during the 1960s and early 1970s that the discipline passed through a phase in which the champions of these two modes of explanation most stridently argued. Insofar as there are elements of uniqueness and generality in all events this is to be regretted; the two modes of explanation are necessarily symbiotic, and until human and physical behaviour is completely predictable (which, at least in the near future, is inconceivable) the complete understanding of an event will involve both idiographic and nomothetic modes of explanation.

Choice between the two modes of explanation is therefore partly one of temperament, and partly dependent upon a personal value-judgement as to which approach is the more useful. This author prefers the nomothetic approach, but in so doing accepts that for the present, and for the foreseeable future, geographical phenomena will not be reducible to an all-embracing set of deterministic laws. However, the search for order in the landscape is considered to be both an interesting and a worthwhile academic pursuit.

In this search for order, it is proper to make use of what has come to be known as the scientific method. In practice two main paradigms can be subsumed under the scientific method, namely the inductive and deductive approaches. These two paradigms have been discussed by Harvey (1969) who queries the existence of the former. They are summarized in Figure 1.1.

Arguments have been made for and against both the inductive and the deductive approaches, but for the present suffice it to say that *in practice* the two approaches are different in degree rather than in kind. For instance, inductive generalizations are not made without some preconceived notions about the processes and relationships involved; indeed these are probably a prerequisite to classifying unordered facts. And conversely, following the deductive approach, one is not likely to produce a valid *a priori* model without some previous experience of the problem area concerned.

The debate on forms of scientific explanation will not be pursued here. The point that has to be made is that, following either paradigm, one reaches a stage where generalizations or hypotheses of one kind or

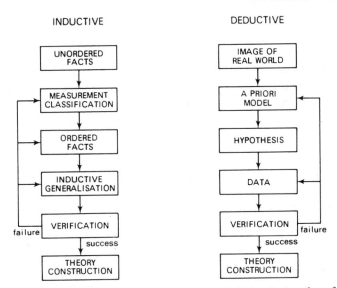

INDUCTIVE

DEDUCTIVE

Figure 1.1: A simplified version of inductive and deductive modes of explanation (modified from Harvey, 1969).

another need to be verified, and one of the most useful ways of verifying hypotheses is to employ statistical methods.

The role of statistics in the social sciences has been made clear by Tukey (1962), and is implicit in the diagrams of the inductive and deductive approaches to explanation presented above. Geographers are not statisticians, and their ultimate objective is not to develop statistical methods: geographers are *data analysts,* and the ultimate objective for those pursuing the nomothetic mode of explanation is the construction of theory. To geographers *qua* data analysts, statistics are merely a means to an end. They are, however, an extremely useful means — witness the burgeoning literature applying inferential statistics to geographic data, and the resulting progress that has been made in building and refining theories of spatial organization.

Descriptive and inferential statistics

Two basic types of statistics may be identified, namely *descriptive* statistics and *inferential* statistics. The descriptive type of statistic is designed to describe a given situation so as to yield quantitative information required by a researcher. A cost-of-living index, for instance, is a descriptive statistic which is designed to keep track of changes in a typical consumer's living costs. Location quotients, pollution counts and evapo-transpiration indices are three descriptive statistics commonly used by geographers.

Unlike descriptive statistics, inferential statistics allow one to make probabilistic statements about the truth of hypotheses, about relationships between variables, and about the population from which a sample is drawn. Consequently, inferential statistics may be used for three specific purposes, namely:

1. To test hypotheses.

This involves setting up a hypothesis, collecting appropriate data, applying a statistical test to the data, and then attempting to decide whether the hypothesis is true or false.

2. To make point estimates.

A point estimate is an estimate of a *population* parameter made on the basis of a *sample.* For instance, we might estimate that the average Australian drinks 68 gallons of beer a year on the basis of the annual consumption of a sample of 1000 Australians.

3. To make interval estimates.

An interval estimate places a confidence interval about a parameter so that for some preassigned percentage of samples the parameter falls within this interval. Thus we might be 95% confident that the average annual per caput beer consumption in Australia lies between 62 and 74 gallons.

To illustrate the difference between descriptive and inferential statistics, imagine we have plotted on a map the distribution of oak trees in a forested area. From this dot map we can compute several statistics which describe the distribution of those dots, one obvious example being the mean distance between the dots. Such descriptive statistics are useful, but they do not convey much information about ecological relationships and the processes influencing the spacing of oak trees. Using inferential statistics, on the other hand, we could examine the hypothesis that oak trees in this forested area tend to be clustered together due to the oak's seed dispersal mechanism, or we could examine the effects upon the spacing of oak trees of soil conditions, slope or other geographical factors. As the title indicates, this book is primarily concerned with inferential statistics.

Levels of measurement

The selection of a statistical test is governed in part by the level at which data are measured. Four ascending levels of measurement have been identified by Stevens (1946). They are as follows:

1. Nominal or classificatory scale.

This is the weakest level of measurement and is used when we can only place observations into mutually exclusive categories. We might, by way of illustration, classify people by their occupation or place of birth. All that is required at this scale of measurement is that members of a sub-class be equivalent.

2. Ranking or ordinal scale.

This is a stronger level of measurement, for the sub-classes can now be ranked above or below one another. In the absence of reliable population data, one might for instance, decide to rank the 12 largest towns in a region by size. The largest town would become number 1, the second largest number 2, and so on down to 12. Sometimes, when ranking data, two or more observations appear to have the same magnitude so that one ends up with observations tying for a given rank. Except in the case of tied rankings, the magnitude of the difference between sequentially ordered observations is always 1.

3. Interval scale.

At the interval scale, not only are observations ordered, but the magnitude of the difference separating any two observations along a measurement scale is variable. Taking the case of city size, differences in the population of any two towns will vary not in terms of rank (as in the case of ordinal data) but in terms of the number of people living in them.

4. The ratio scale.

The ratio scale is the highest level of measurement, in which measurements are made relative to some absolute and non-arbitrary base. For example, if we measure the distance from two farms, A and B, from a creamery both in miles and kilometres, we would find that the ratio of the distances is identical using both measuring systems: if farm A is located three times as far from the creamery than is farm B in miles, then it is located three times as far using the metric system. This clearly does not apply to all measurement scales. For instance, given two weather stations, one recording 8°C and the other 16°C, then the same ratio, 1 : 2, would not apply if these temperatures were converted to the Fahrenheit scale. In passing it may be noted that some people argue that there is no true origin for any phenomenon, therefore the ratio scale is purely hypothetical. While this may be true it does not affect the statistics to be presented in this book; therefore we will side-step the issue.

Levels of measurement are introduced at this point because they have some bearing upon the selection of a statistical test.

General requirements for a statistical test

Certain minimum requirements need to be met before it is possible for a geographer to consider the question of which test to apply. The main requirements are:

1. Data must be available with which to test a hypothesis.
One cannot make meaningful inferences using hypothetical data; hence if a geographer wishes to investigate a specific problem, it is often worth-while to begin by exploring data sources.

2. Data must be reasonably reliable.
It is very rare for data to be minutely accurate, and even such exalted sources as national censuses can be surprisingly inaccurate. To take an extreme case, the number of 20–24-year-old males in the 1962 census of Kenya appears to be approximately 50% inaccurate, and certainly grossly underrecorded, as a result of the political unrest at that time. Error creeps into data as a result of deliberate misreporting, carelessness and a whole host of other reasons.

3. A sample must be large enough for the purposes at hand.
For instance the one sample chi square test based on two categories discussed in Chapter 6 needs a minimum of ten observations to be reliable. In designing an experiment it is important to collect a sample at least as large as is required by the statistical tests to be applied; preferably it should be larger for, other things being equal, the larger the sample the easier it is to make a decision to reject or accept a hypothesis.

4. One ought to specify the correct sampling distribution (page 30) for a statistical test, which is difficult when one is faced with incomplete information about complex processes. For example are mononucleosis and leukemia noncontagious, or are they susceptible to low-level epi-demics which diffuse through space? If the former were true, then one would use the Poisson distribution, but if the latter condition prevailed then Neyman Type A, or some similar distribution, might be more appropriate.

A parametric or a nonparametric test?

The late Sidney Siegel, who was one of the most ardent advocates of nonparametric methods, stressed several advantages of nonparametric tests in his book *Nonparametric Statistics for the Behavioural Sciences*

(1956). Five of his more potent reasons are:

1. They are distribution free and do not assume that a sample is drawn from a population which is normally distributed and homoscedastic (this term will be explained in a later chapter).

2. They may be used with data measured at the ordinal or even the nominal scale of measurement.

3. They are easy to compute.

4. They are useful with small samples.

5. Many nonparametric tests are exact tests, i.e. they give rise to exact probability statements regardless of the distribution of the population from which the sample is randomly drawn. It is stressed, however, that quite a few nonparametric tests are approximate tests; Sawrey (1958) refers to these approximate tests as semi-nonparametric tests.

Reason 5 does not appear to be open to dispute. However the first four reasons deserve examination, for research in recent years suggest that Siegel may have overstated the case for nonparametric methods.

Reason 1 implies that parametric tests can only be used when the underlying population from which the sample is drawn is normally distributed and homoscedastic. However Donaldson (1968) has found empirically that the F test (one of the more important parametric tests) is quite 'robust' when applied to non-normal data, and to data with unequal variances. Robustness means that a technique is relatively insensitive to violation of its assumptions, so that a hypothesis is accepted when it should be accepted, and rejected when it should be rejected. Donaldson also found that when the assumptions were violated, in most situations the F test tended to be 'conservative', which may be loosely interpreted to mean that the test errs on the side of caution and establishes significance slightly less frequently than it ought. Conversely a 'liberal' test establishes significance slightly more frequently than it ought. (The terms 'conservative' and 'liberal' are normally used with respect to Type I and Type II errors which are discussed in Chapter 2.) It would seem, therefore, that Siegel's assumption that data must be normally distributed and homoscedastic if a parametric test is to be applied is not wholly true. Two of the tests presented later, the F and the related t test, work fairly well when the two assumptions mentioned above are moderately violated, although in cases of extreme deviation a nonparametric test is more appropriate.

Reason 2 for using nonparametric methods implies (as Siegel (p. 19) states explicitly with reference to the t test) that parametric tests require measurement in at least the interval scale. Anderson (1961) on the other hand states categorically that Siegel's statement is completely incorrect, and describes an application of a parametric test to data measured at the ordinal scale. As a general rule, nonparametric tests are applied to data measured at the nominal and ordinal scale, and parametric tests to data measured at the interval and ratio scale. However there are instances in which parametric tests may be applied to data measured at the ordinal scale. Nonparametric tests may, of course, be applied to data measured at the interval and ratio scale by reducing the data to a lower level of measurement, but in practice this may involve throwing away valuable information.

Reason 3: writing in 1956 it is understandable that Siegel would find nonparametric tests easier to compute. However the subsequent development of computer science throws this reason open to debate. Many parametric methods are now available as packaged computer programs; hence to run a test all one needs to do is punch the data on computer cards in the required manner or type it in at a remote terminal. The result has been to speed up the time of computation. Likewise, some nonparametric tests have been programmed, and several packaged programs permitting a variety of nonparametric tests are now available. On the whole, the effect has been to narrow the gap, in terms of computational simplicity, between parametric and nonparametric tests in those places where computers with the appropriate software are accessible. Beyond the 'computershed', however, nonparametric tests do still have this advantage of computational simplicity.

Reason 4 has been disputed by Gaito (1959), who states (pp. 119–20):

Statements by Box and Anderson, and others . . . indicate that as long as the distribution does not deviate too greatly from normality, an investigator may use a parametric technique no matter what the size of the sample. . . . The unreliability of small samples is considered in the tabular values required for significance.

The problem in conducting tests with small samples would appear to be not that the tests are inapplicable, but that it is difficult to know whether one is meeting the assumptions of a test or using the correct sampling distribution. With six observations, for example, one cannot tell whether the populations from which a sample is selected is normally distributed or highly skewed, whereas with a sample of thirty observations a fairly clear picture emerges. The solution lies in one's knowledge of the population or process being examined: if there is good evidence from parallel examples that a test is applicable, then one may be reasonably confident about applying it to a similar situation with a small sample.

One further, and quite different, argument in favour of non-parametric methods will be added: they are normally easier to understand and hence have certain advantages for instructional purposes. It is easier to grasp such concepts as the role of chance or random events, confidence levels, and the principles of hypothesis testing in the context of nonparametric tests, and then, having mastered this much, to advance to the mathematically more complex parametric tests. This is the approach used in this book: for pedagogic reasons, non-parametric methods are dealt with in Part two and parametric methods in Part three.

The power of a test

One other important consideration influencing the choice between a parametric and a nonparametric test is the power of a test. The power of a test is defined as the probability of rejecting the null hypothesis when it is true. This definition will be elaborated in Chapter 2 when Type I and Type II errors are discussed, but for the present suffice it to say that this is exactly what we want a statistical test to do. Insofar as parametric tests are usually more powerful than equivalent nonparametric tests, this implies that where a choice exists parametric tests have the edge over nonparametric tests.

Siegel (1956, p. 20) counters this argument by pointing out that if the size of a sample is increased, a test becomes more powerful: the term *power-efficiency* is used to describe the increase in sample size that is needed to make one test as powerful as another. Hence if we have a parametric test A and a nonparametric test B, the latter can be made more powerful than the former by using an enlarged sample with test B. This would seem to be a spurious argument, for if the sample size can be increased for test B then it can also be increased for test A, and the parametric test would still be more powerful for a sample of a given size.

It might appear that one will often be faced with a choice amongst a bewildering variety of statistical tests. In reality, the choice is usually curtailed to a point where very few possibilities exist; indeed quite often only a single test may apply. Where a choice does exist, para-metric tests are usually to be preferred because they are more powerful, provided that their various assumptions are met.

2 Conducting a statistical test

Three branches of statistical inference were identified in Chapter 1, namely hypothesis testing, point estimation and interval estimation. Insofar as this book is concerned mainly with the first of these, various considerations involved in conducting a statistical test warrant discussion. Statistical tests are based on samples, which is why inferential statistics are sometimes referred to as 'sample statistics'. Three aspects of samples will be examined: the distinction between a population and a sample; problems involved in selecting a sample; and the importance of a sampling distribution. Topics leading up to the presentation of a standard procedure for conducting a test include discussion of the null hypothesis, one- and two-tailed tests, levels of significance, and Type I and Type II errors. Finally, some common sources of error in statistical tests are reviewed.

Population and sample

A *population* is defined as a universe comprising *all* members of a specified group. A population may be finite or infinite. For instance the population consisting of all university students in Canada is finite, while the population of outcomes (heads or tails) from tossing a coin is infinite.

A *sample* is a part of a given population, and is invariably of a finite size. If we define a population as a set, then a sample is a sub-set drawn exclusively from that set, but consisting of fewer individuals than the number in the set as a whole. Thus a sample might consist of all the students attending Canadian universities born in a given month, or thirty tosses of a coin. Not only do samples vary enormously in absolute size, but also they vary from very small proportions of a population (as in consumer surveys and public opinion polls) to very large proportions (as in plebiscites and general elections).

The two types of questions most commonly found in hypothesis testing are:

1. What is the probability that two (or more) samples are drawn

from different populations?

2. What is the probability that a given sample is drawn from a population which has certain defined characteristics?

The definition of a population has received some attention in the geographical literature in the last few years. Both Gould (1970) and Meyer (1972) have suggested that geographers frequently make the mistake of using populations as a basis for statistical inference when, correctly, samples should be used. Court (1972) challenges Meyer's position: since statistical populations are idealizations, it is perfectly proper to apply inferential statistics to what Meyer would call a population. Cliff (1973) has contributed to this debate by demonstrating that inferences may be drawn from data which might appear to represent a population if either the results appear to be unusual in some way (in comparison with all possible random permutations of the data), or if a stochastic (random) process underlies the observed distribution. The position adopted here is that since we can almost invariably conceptualize an ideal population about which inferences can be made, and since random processes affect nearly everything we observe, it is quite reasonable to treat most geographical data sets as if they were samples and to apply inferential statistics. In other words, the 'real world' is an empirical realization amongst an infinite number of hypothetical realizations of a particular generating process. By applying inferential statistics to empirical data, we hope to learn something about the generating process. It is of interest that the sociologist, Blalock (1960, p. 270) has reached a similar conclusion in stating: 'some sort of significance test will practically always be helpful in evaluating one's findings'.

Selecting a sample

Two general types of sampling procedure may be identified, *purposive* or *subjective* sampling, and *probability* or *objective* sampling. Purposive samples are arbitrarily and subjectively chosen by a researcher. Perhaps the commonest purposive sample used by geographers is the case study. Thus a particular farm or river basin is selected as a case study for detailed discussion and analysis. Street interviews when an interviewer has to fill a quota is another type of purposive sample. Two problems can arise from using purposive samples in statistical analysis. First, it is possible for two experts to select quite different samples, leading to conflicting conclusions. Second, many purposive samples are not typical, but are ideal or perfect examples; these are likely to be biased, and therefore inappropriate for statistical analysis.

It is not intended to imply that all purposive samples are bad samples; indeed they can, in some situations, be relatively unbiased and precise. For instance, in a study of residential mobility, Adams (1969) selected from the Minneapolis street directories as a sample of movers all those whose surname began with the letter K because this provided a manageable number of observations, and K was the first letter in the alphabet for which there was no pronounced ethnic bias by surnames. As a general rule, however, statistical inference is better served by probability samples than by purposive samples.

Probability samples are random samples drawn in such a way that, beyond picking the sampling design, the preferences of the researcher do not influence the choice of individuals for inclusion in the sample. Geographers encounter two types of sampling situations in which probability sampling methods are useful: the first involves objects, the second, areas.

Suppose one is studying land purchases over a five-year period in rural areas surrounding a metropolis such as Paris. In this case the 'objects' being studied are parcels of land that have changed ownership. Suppose 990 transactions have taken place, i.e. the total population is large, and that analysis is to be based on a sample of these transactions. Two simple methods of sampling might be used. The first involves using a random number table. These tables (which are to be found in many statistical textbooks) consist of blocks of numbers that meet certain properties of randomness, the most important of which are: that numbers in the range 0 to 9 are equally likely to occur; and that the numbers are not serially ordered in any way. Starting at any point in the table, one moves systematically through the table taking, in this example, 3 digits at a time. If the number selected falls in the range 001 to 990, the transaction represented by that random number is included in the sample. Ignoring numbers that occur more than once, one continues to select from the random number table until one has a *random sample* of the required size. The second method produces a *systematic sample:* one begins from a random starting point, and selects individuals from there on using a convenient fixed lag. In the above example, one might begin with the 7th transaction, and then select every tenth one up to the 987th transaction. Systematic samples should not be used if the data are ordered in any way that will influence the characteristics of the resulting sample — for instance no periodicities should be present in the data.

More elaborate methods of sampling include *multistage* or *hierarchical* designs in which the population is first stratified, certain strata are randomly selected, and then within the strata the individuals are also randomly selected.

In sampling areas (the correct term is 'plane sampling') geographers are faced with a special problem, for there are an infinite number of points in space. Three methods of plane sampling may be identified: point samples, line samples or traverses, and areal or quadrat samples. The literature is mainly concerned with point samples. A variety of point sampling designs exist, the majority being extensions and modifications of the random, stratified, and hierarchical designs discussed above. The basis of plane sampling is that two numbers are needed to identify a point, one for the northing, the other for the easting. In random plane sampling one selects a pair of random numbers that represent the geographical coordinates of a point, and notes the appropriate information — such as land use — at that point. Further pairs of random numbers are selected until the sample is of the required size. It is quite common with this sort of random sampling for certain areas in the study region to have quite a cluster of sampling points, and for other areas to be unsampled. Some researchers, wanting a more even distribution of sampling points, use a stratified plane sample: the study area is covered by a grid, and one or more sampling points are randomly selected within each quadrat defined by the grid. Other variations, including aligned and unaligned sampling designs are well reviewed in Berry and Baker (1968) and Harvey (1969, pp. 356—69).

The quality of a sample is assessed in terms of the following three criteria:

1. Bias.
A good sample is unbiased so that estimates of population parameters tend to be neither systematically larger nor systematically smaller than the true value. Of course a single estimate may be higher or lower than the true value, but the estimate of a parameter obtained from repeated sampling ought to be identical with the true value of that parameter.

2. Precision.
A good sample provides an estimate of a population parameter that is as close as possible to the correct one.

3. Ease of collection.
The collection of a sample involves inputs of time, manpower, and money, hence if two sampling designs provide equally unbiased and precise estimates, then the design which makes the sample easier to collect is the better.

In principle, it is important that the statistical criteria should override the convenience criterion. This ideal should be adhered to even when a ready-made sample is available if there are grounds for supposing that the sample is biased or unrepresentative.

The sampling distribution

The sampling distribution is a theoretical distribution: it is defined as *the distribution that would be obtained by randomly drawing ALL POSSIBLE SAMPLES of a given size from a specified population.* The simplest example is provided by coin-tossing. Suppose we toss a coin three times and get three tails: what is the probability of this outcome? First let us find the sampling distribution resulting from three tosses of a fair coin. The list of all possible combinations is given in Table 2.1.

Table 2.1   All possible outcomes with three tosses of a coin.

	\multicolumn{8}{c}{Outcome}
	1	2	3	4	5	6	7	8
Toss 1	H	H	H	T	T	T	H	T
Toss 2	H	H	T	H	T	H	T	T
Toss 3	H	T	H	H	H	T	T	T
Total number of tails	0	1	1	1	2	2	2	3

A total of eight possible outcomes exists — and only one of these (the eighth outcome) is the observed combination of three tails: hence we may say that the probability of three tails, under random expectations, is 1 in 8, or $P(3T) = 1/8 = \cdot 125$, (where P denotes the probability of an event occurring).

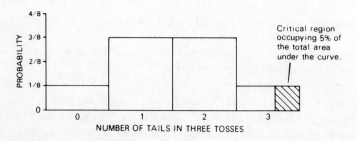

Figure 2.1: Histogram showing the sampling distribution for three tosses of a coin.

The sampling distribution in Figure 2.1 gives the probability of all possible outcomes: for example there are three ways of obtaining two tails, namely outcomes 5, 6, and 7, hence the probability of two tails is 3/8 or $P(2T) = \cdot375$, and so on for one and zero tails.

While it is easy to work out the sampling distribution in a simple case like three tosses of a coin, in most cases it is not feasible to work out the distribution of all possible outcomes because the number of combinations is so large. For example, if a six-sided dice is tossed five times there are 7776 possible combinations. We therefore usually fall back on mathematical theorems which make use of known distributions: these provide either a close approximation to the presumed distribution, or in some cases, an exact probability distribution.

Once the sampling distribution of a given statistic is known, one can then make probabilistic statements about a sample so as to test hypotheses.

The null hypothesis

Closely associated with the concept of a sampling distribution is the *null hypothesis* (which is represented by the symbol H_0). In fact we may state that the sampling distribution of a statistic records the probability associated with a given numerical outcome under the null hypothesis. Siegel (1956, p. 7) defines the null hypothesis as a hypothesis of 'no difference'. This succinct definition deserves some elaboration by way of illustration. Say we were to compare the intelligence of male geography students to that of female geography students as measured by an IQ test. Our normal procedure is to set up a null hypothesis stating that there is no difference between the average IQ of male and of female geography students: one then proceeds to try to reject this hypothesis of no difference. Clearly, not every male and female student has an identical IQ; there are considerable differences from individual to individual so that if one randomly picked ten male and ten female students, the sample means could be quite different even when the population means (for all the male and female and students) are identical. This would happen if, by chance, one picked ten rather bright females and ten rather dull males. Such chance or *stochastic* sampling effects are explicitly built into statistical tests, as will become clearer when levels of significance are discussed.

Normally the null hypothesis, H_0, is compared to the *research* or *alternate* hypothesis, H_1. It is commonly, but not invariably, the case in statistical analysis that we wish to reject the null hypothesis so as accept the alternate hypothesis. There are instances, however, when the null hypothesis is set up in the hope that it is true.

Several statisticians are sceptical about the use of the null hypothesis in decision-making. Rozeboom (1960) 'vigorously excoriates' the null hypothesis decision procedure as a method of inference, particularly because it appears to involve a definitive decision, whereas it ought to be part of a cognitive (learning) process. In other words one should not treat rejection of the null hypothesis as a conclusive decision implying that the alternate hypothesis is 'true': rather, rejection should be viewed as a piece of evidence increasing one's degree of belief in the truth of the alternate hypothesis.

Three remarks can be made concerning this criticism by Rozeboom. First, it seems to imply that people treat hypotheses rather like a defendant on trial who pleads not guilty: he is either found to be telling the truth, and is accepted back into society, or is found to have perjured and is cast asunder. Yet in hypothesis testing things are not so clear-cut. Neyman (1953) points out that tests should only be a guide to behaviour, so that acceptance of a hypothesis does not require us to believe that the hypothesis is true. Second, the situation Rozeboom describes does not appear to be so much that of a student conducting a series of loosely related tests, as of a researcher making a series of exhaustive and fairly repetitive tests of a narrowly defined problem. Also, it does not reflect the non-experimental nature of geography. As a psychologist, Rozeboom is often in a position to repeat an experiment if it produces inconclusive results at first. For the most part, geographers observe situations that cannot be reproduced experimentally: hence we have to live with inconclusive results, and try to decide whether they have implications for theory construction, or should be ignored. Third, even where repeated experiments are feasible, at some point a statistician has to make a decision about a hypothesis. It has to be remembered that statisticians work in probabilities and there is no such thing as 100% certainty. Although there are times when the evidence is overwhelming, more often the results of a sequence of experiments are less than conclusive, and sometimes contradictory, whereupon one is still faced with having to make a decision which is, at least in part, arbitrary.

In summary, it is suggested that the null hypothesis decision procedure is a useful device, provided that it is not elevated to a ritual status. Obviously the researcher must be flexible, and especially when he finds a hypothesis is barely rejected, or marginally accepted, for in such a situation a small change in one's level of significance leads to a completely different decision. As emphasized in Chapter 1, statistics are to the data analyst but an intermediate step in the search for explanations: if that search is to be conclusive, some sort of decision procedure is needed lest the means prevent the achievement of the final goal. Subject to the caveats discussed above, the null hypothesis procedure constitutes a satisfactory basis for decision-making.

Levels of significance

In rejecting or accepting a null hypothesis, we cannot be certain that we have made a correct decision, since statisticians work in terms of probabilities rather than certainties. An integral part of the null hypothesis decision procedure is the *level of significance* at which null hypotheses are to be accepted or rejected. The level of significance (denoted by the Greek letter α — alpha) is specified before a test is conducted.

It is common to make α small, so that some confidence can be placed in the result: it would not be much help, for instance, to set α at ·5, for that would imply that half of the time one could be making the wrong decision in rejecting the null hypothesis. By convention, the two most commonly used significance levels are:

1. α = ·05, meaning that the probability of rejecting the null hypothesis when it is true is 5 times in 100, or 1 in 20 times.

2. α = ·01, meaning that the chances are that 1 in 100 times one would be incorrect in rejecting the null hypothesis.

It is perfectly legitimate to use other levels of significance. Choice of α should be governed by the utility of the decision to be made: where the decision is of critical importance, α may be set as high as ·0001, but in the majority of situations $\alpha_{.05}$ or $\alpha_{.01}$ are used. In some cases two significance levels are applied, so that tests are conducted at $\alpha_{.05}$ *and* at $\alpha_{.01}$: this creates three categories of relationships which are commonly referred to as 'non-significant', 'significant', and 'highly significant'.

Figure 2.1 shows the sampling distribution for three tosses of a fair coin; as in all sampling distributions, the total area of the distribution, when plotted, corresponds to a probability of 1·0. Within this total area there is a *critical region*, the region in which one is critical of the null hypothesis. The boundary of the critical region is defined by the significance level that has been adopted. Suppose that we are examining the likelihood of obtaining three tails in three tosses of a fair coin, and that α = ·05. The hypotheses are as follows:

H_0: the coin is not biased and the observed frequency of tails differs from the frequency expected using a fair coin only by chance.

H_1: the coin is biased towards tails so that the frequency of tails is higher than would be recorded by a fair coin.

From Figure 2.1 it will be seen that the outcome of three tails occupies the upper end of the sampling distribution, and has an area (probability) of ·125. The critical region is even smaller than this, and occupies only a fraction of the area representing the outcome three tails. Thus in this example, the observed outcome (3T) does not lie *wholly* within the critical region, we cannot reject the null hypothesis, and we conclude that there is not sufficient evidence to validate the hypothesis that the coin is biased towards tails.

In some books the term 'confidence level' is used in place of 'level of significance'. In the context of hypothesis testing, these two terms can be used interchangeably so that the 95% confidence level is, for practical purposes, the same as the ·05 significance level, and the 99% confidence level is the same as the ·01 significance level, and so on.

The term '95% confidence level' indicates that one is making the correct decision 95 out of 100 times in rejecting the null hypothesis when an observation or outcome lies wholly within the critical region. The term '·05 significance level' implies that, under the null hypothesis, only 5 times in 100 will an observation be so significantly different from expectations that it occupies the critical region, i.e., only 5 times in 100 will one make the error of rejecting a null hypothesis when it is true.

One- and two-tailed tests

A distinction can be drawn between those statistical tests which are one-tailed, and those which are two-tailed. The phrasing of the alternate hypothesis, H_1, determines whether a test is one- or two-tailed. In the case of a *directional* hypothesis such as '$A > B$', or 'X is positively correlated with Y', a one-tailed test is applied. In the case of a hypothesis merely stating *difference,* for instance '$P \neq Q$', or 'X is significantly different from Y', a two-tailed test is used.

The location of critical regions for one- and two-tailed tests are shown in Figures 2.2 and 2.3 respectively. In the one-tailed case, the critical region is located in one of the tails, whichever is specified by H_1, and if $\alpha = ·05$ then the critical region occupies 5% of the total area under the curve describing the sampling distribution. In the two-tailed case, there is a critical region in *both* tails, and α is divided equally between the two

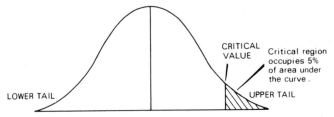

Figure 2.2: A one-tailed test based on the normal distribution with the critical region in the upper tail (α = ·05).

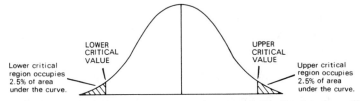

Figure 2.3: A two-tailed test based on the normal distribution (α = ·05).

critical regions: this means that when α = ·05, the upper and the lower critical regions both occupy 2·5% of the area under the curve.

One aspect of the null hypothesis decision procedure which often appears to be arbitrary is the choice between a one- and a two-tailed test. Simply by switching from a test of direction to a test of difference, and retaining the same level of significance, the area in the critical region in one tail is halved, while a new critical region is created in the other tail which previously lacked a critical region. However the arbitrariness of this difference is more apparent than real: if the hypothesis being tested is soundly formulated, then it should be quite apparent that either a one- or a two-tailed test is required.

Type I and Type II errors

These two types of errors are defined as follows:

A Type I error is to reject the null hypothesis when it is true. This may be characterized as 'a sin of commission' for a significant relationship is being discovered when none really exists.

A Type II error is to accept the null hypothesis when it is false. This may be characterized as 'a sin of omission' for one has failed to identify a significant relationship where one actually exists.

From these definitions, it is clear that the former error is, in most circumstances, more serious than is the latter.

What are the probabilities of making one of these errors? Using P to denote the probability of something happening, then

$$P \text{ (Type I error)} = \alpha.$$
$$P \text{ (Type II error)} = \beta.$$

In other words in choosing a given level of significance, one also determines the probability of making a Type I error. It is the seriousness of Type I errors that leads statisticians to set demanding levels of significance.

Type II errors are connected with the discussion in the previous chapter on parametric versus nonparametric tests. The power of a test is given by

$$\text{power} = 1 - P \text{ (Type II error)} = 1 - \beta.$$

Hence the greater the ability to identify a significant relationship, the more powerful is a test. Parametric tests are more powerful than the equivalent nonparametric tests because they are more efficient at accepting research hypotheses that are true.

The terms 'conservative' and 'liberal' which were alluded to in Chapter 1 are normally used with reference to Type I and Type II errors. If a test is defined with α level of protection against a Type I error, then even more protection against a Type I error exists when the test is conservative. Thus if one were conducting a test and $\alpha = \cdot 05$, in the case of a conservative test, the null hypothesis is incorrectly rejected less than 5 times in 100, whereas in the case of a liberal test, the null hypothesis is incorrectly rejected more than 5% of the time. Likewise for Type II errors: a conservative test commits an error of the second type less frequently than an equivalent liberal test, so that conservative tests, according to the definition above, are more powerful.

A simple procedure for conducting a statistical test

The test procedure outlined below, and used in subsequent chapters, may appear to be highly mechanistic, but it is convenient for purposes of presentation. In practice it is perfectly correct to modify the procedure to suit particular situations.
The four steps used here are:

STEP 1: PROBLEM DEFINITION
(a) State the null hypothesis (H_0) and the research hypothesis (H_1).

If the case arises, make explicit whether the test is one- or two-tailed.
(b) Identify an appropriate level of significance, α.
STEP 2: TEST SELECTION
(a) Determine whether the data are measured at the nominal, ordinal interval or ratio scale.
(b) Determine the sample size (N).
(c) Select an appropriate statistical test.
STEP 3: DETERMINE THE SAMPLING DISTRIBUTION
If computation is required, evaluate either the whole of the sampling distribution, or simply the tails of the sampling distribution, to obtain the critical value: if the sampling distribution is tabled, look up the critical value in the appropriate table.
STEP 4: COMPUTATION
Compute the various statistics, and on the basis of this accept or reject H_0.

Some common sources of error in statistical analysis

Even the greatest statisticians make mistakes. Students may rest assured, therefore, that in making errors (as they will surely do) they are in distinguished company. It should be noted, however, that many errors are avoidable; by listing some of the more common sources of error, their incidence may be reduced:

1. Arithmetic errors. These are the most common of all, and no doubt will persist until the end of time. One cannot overstress the value of checking one's calculations, or where mathematical checks are available, making use of these checks, however tedious this may seem.

2. Incorrect numerical precision. There is no virtue in making some calculations using twelve decimal places, and then rounding the result to an integer. Likewise there is no virtue in excessive rounding, and then calculating some division to several significant figures: the result would be spuriously accurate. Two general rules are suggested: first, work with a number of digits appropriate to a given problem — neither over-generalize nor be excessively accurate. Second, be consistent — try to maintain the same level of accuracy throughout all the computations.

3. Poor sample selection. A sample is supposed to provide an accurate picture of the population which it purports to represent, and for this reason a variety of sampling designs have been developed, as discussed early in this chapter. Unfortunately, all too often good samples are not available, and the results have to be interpreted with due caution.

4. Frequent Type I errors. These occur when the significance level, α, is set too low.

5. Improper treatment of extreme observations. Here one is in a dilemma, and the solution has to be highly subjective. For instance, how would one treat a report of a person being 8 feet 7¼ inches tall? *The Guinness Book of Records* provides grounds for supposing this to be an error — perhaps 6 feet accidently transcribed as 8 feet. On the other hand a radio operator, on hearing of the Pearl Harbor attack on 7 December 1941, decided it was a preposterous hoax and did not bother to relay a message about it: not all extreme events are untrue. Although it is correct to suspect anomalous measurements, great care must be taken in rejecting observations simply because they are suspicious, for they can have a big influence on several sample parameters. Where possible one should trace an observation back to its original source and check there.

3 Central tendency, dispersion and the moments of a distribution

Later chapters of this book assume knowledge of certain preliminaries which are discussed in this and the following chapter. This chapter deals with the various parameters that describe a distribution, namely measures of central tendency, of dispersion, and other higher order moments. These descriptive statistics are referred to in a number of the nonparametric tests, but they have greatest relevance to the parametric section. The reason for this is that, as the name implies, parametric tests make assumptions about the parameters describing a distribution.

Measures of central tendency

The terms 'central tendency' and 'average' are used synonymously, hence each of the five measures of central tendency discussed below — the mode, median, arithmetic mean, geometric mean and harmonic mean — provides a different way of measuring the average for a set of observations.

The simplest measures of the average are computed for one-dimensional series in which the observations, be they ages in years or incomes in roubles, lie along a single linear scale. Averages may also be computed for two-dimensional or *bivariate* distributions, and for multi-dimensional data. Geographers make frequent use of one-dimensional variables, but there are also instances when one wishes to find the 'centre' or average position of a two-dimensional spatial distribution which is, in effect, a bivariate distribution. Where, for instance, is the population centre of Australia? In this case each observation is measured on two scales, a northing (giving the position north or south of an arbitrary origin) and an easting (giving the corresponding position of an east–west scale). Accordingly, the mode, median, and arithmetic mean will be considered with reference to both one- and two-dimensional series.

The study of central tendency in two-dimensional distributions has been the concern of a group who use the term 'centrography' to describe their work. Centrography flourished in the 1920s and 1930s when, for instance, there existed in Leningrad the Mendeleev Centrographical Laboratory. This approach has since been given little prominence, although Stewart and Warntz and other people associated with the American Geographical Society Macrogeography Project in the late 1950s and early 1960s have made a strong case for reviving centrography in a revised form so that it is integrated into the mainstream of geographical thinking. The brief discussions of centrographic measures that are presented below are based largely on the work of D. S. Neft (1966) who worked on the Macrogeography Project at that time.

The mode

Suppose that the temperature at a weather station is recorded in degrees centigrade at noon on nine consecutive days. If we call the temperature variable X, then we have a vector, X, composed of nine elements, the measurements, as follows:

$$7\cdot3 \quad 10\cdot7 \quad 9\cdot1 \quad 8\cdot4 \quad 13\cdot9 \quad 9\cdot4 \quad 6\cdot1 \quad 10\cdot8 \quad 8\cdot9$$

The mode is the most frequently occuring value in an array of numbers; in X, above, no value occurs more than once, so that we might conclude that no modal value can be identified. More commonly, with this kind of data, the observations are arranged into classes. Doing this, and making sure the class intervals are of equal size, we obtain:

Class interval	Observations	Frequency
$6\cdot0°-7\cdot9°$	$6\cdot1°$ $7\cdot3°$	2
$8\cdot0°-9\cdot9°$	$8\cdot4°$ $8\cdot9°$ $9\cdot1°$ $9\cdot4°$	4
$10\cdot0°-11\cdot9°$	$10\cdot7°$ $10\cdot8°$	2
$12\cdot0°-13\cdot9°$	$13\cdot9°$	1

Clearly, when the data are arranged into classes, a modal class ($8\cdot0°-9\cdot9°$) can be identified and if the grouped data are plotted as a histogram, as in Figure 3.1 the modal class stands out as the peak in the distribution. A mode, or a modal class, may or may not exist, as follows:

1. When there is a single dominant frequency, the distribution is *unimodal*.

2. When there are two equally dominant frequency the distribution is *bimodal*.

3. In cases where several modes or modal classes occur, then one might be inclined to describe such a distribution as *multimodal*: however one soon reaches a position where it is more realistic to say that no mode exists.

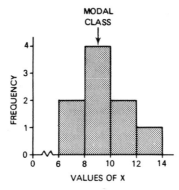

Figure 3.1: Histogram of variable X.

The mode is an appropriate measure of central tendency in situations where one is catering primarily to the needs of the most populous category in a distribution. For instance, suppose one were planning a housing estate, and a local by-law stipulated that each child must be provided with a separate bedroom. Given that the 'average' family has 2·37 children, then it would be ludicrous to plan an estate of houses with 3·37 bedrooms. If the modal family had 2 children, then most of the houses would have 3 bedrooms, and the purchasing of building materials, the layout of house lots and so on would be largely tailored to the needs of this modal class.

The two-dimensional measure corresponding to the mode is known as the modal centre (MO_c). The modal centre of an areal distribution is that unit of area occupied by the greatest number of individuals, i.e., the most densely populated area. For instance, the modal centre of Australia, as shown in Figure 3.2, is Sydney in New South Wales.

The median

The median divides a set of ranked observations into two groups of equal size. If one has N ranked observations, then the position of the median (M) is given by

$$\text{position of } M = \frac{N}{2} + \frac{1}{2}$$
    3.1

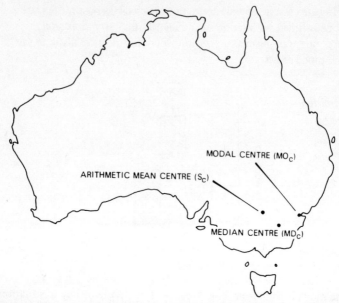

Figure 3.2: Centrographic measures for Australia (based on Neft, 1966, p. 41).

For the 9 temperature observations, M occupies the fifth position, and the fifth observation, counting from either smallest to largest, or largest to smallest, is $9 \cdot 1°$. When there is an even number of elements in an array, then a unique median value cannot be identified; in this case the median is arbitrarily defined as the mid-point between the two values about the median position. For instance in the case of an array of 12 numbers, if the 6th ranked value = $4 \cdot 7$ and the 7th ranked value = $5 \cdot 5$, then the median is

$$\frac{(4 \cdot 7 + 5 \cdot 5)}{2} = 5 \cdot 1.$$

The median has the property that it divides the area under a frequency distribution in half, so that one can state whether a given individual belongs in either the lower or upper half of a frequency distribution. It follows that the median is an appropriate measure of central tendency for ordinal scale data.

The median has another important property. Suppose that one picked an arbitrary value, P, in the range of variable X; when P equals the median, the sum of the absolute deviations about P are at a minimum. Hence

$$\Sigma \ |X_i - P| = \text{minimum when } P = M. \qquad \textbf{3.2}$$

(The vertical lines in **3.2** represent the 'absolute value', i.e., negative signs are ignored.)

The generalization of this principle to two dimensions bestows upon the corresponding centrographic measure, known as the *median centre* (MD_c) a useful property: it is the point of *minimum aggregate travel.* Because of this property, the median centre has many potential applications. For instance, suppose one wished to locate a school in a rural area, and the distribution of the pupils were known: the median centre would, at least in theory, minimize the total distance travelled by the pupils. In the Australian example (Figure 3.2), the median centre lies approximately half way between Sydney and Melbourne.

The arithmetic mean

The arithmetic mean is given by the sum of an array of numbers divided by the total number of observations in that array. The mean of a sample is normally represented by placing a bar above the name for a variable. Thus $\bar{X}$ (called 'X bar') is obtained from

$$\bar{X} = \frac{\sum\limits_{i=1}^{N} X_i}{N} \qquad \textbf{3.3}$$

In the case of the temperature data, we have

$$\bar{X} = \frac{7\cdot3 + 10\cdot7 + 9\cdot1 + 8\cdot4 + 13\cdot9 + 9\cdot4 + 6\cdot1 + 10\cdot8 + 8\cdot9}{9}$$

$$= \frac{84\cdot6}{9} = 9\cdot4$$

The median for this set of data is $9\cdot1$, hence the arithmetic mean (or more simply, the mean) is slightly larger. The relationship between the mode, median and mean will be examined more closely later in this chapter when skewness is discussed.

The mean has two interesting properties. As the median is the value about which absolute deviations are minimized, so the arithmetic mean is the quantity about which squared deviations are minimized. Hence for any value, P

$$\Sigma \ (X_i - P)^2 = \text{minimum when } P = \bar{X} \qquad \textbf{3.4}$$

The second property states that the sum of deviations from the mean of an array equals zero. Using lower-case x to represent a deviation score (i.e., $X_i - \bar{X} = x_i$), we have

$$\sum_{i=1}^{N} (X_i - \bar{X}) = \Sigma\ x_i = 0 \qquad\qquad \textbf{3.5}$$

It follows that the average deviation about the mean also equals zero, hence

$$\frac{\Sigma(X_i - \bar{X})}{N} = \frac{\Sigma x_i}{N} = 0 \qquad\qquad \textbf{3.6}$$

This relationship has a proper statistical name: it is the *first moment about the mean* or, more briefly, the *first moment,* and we can state axiomatically that the first moment always equals zero.

The arithmetic mean centre (S_c) is the centre of gravity of a distribution: it is the point where the square of the deviations is minimized, and hence is more sensitive to extreme values than is the median centre. This shows in the map of Australia where the mean centre lies further toward the interior of the continent than does the median centre.

The geometric mean

This measure of central tendency is not encountered very frequently. It is appropriate to finding the mean of a series that tends to grow geometrically, and hence is used to average ratios. Using G to denote the geometric mean, then

$$G = \sqrt[N]{X_1 \times X_2 \times X_3 \times \ldots \times X_N} = \sqrt[N]{\prod_i X_i} \qquad\qquad \textbf{3.7}$$

One useful application of this measure in geography is in estimating the population of urban centres for years mid-way between censuses. It is not unreasonable, in the absence of other information, to assume that the population of an ubran centre grows geometrically. Suppose a village had 100 inhabitants in 1880, and 200 in 1900; then the estimate for 1890 is

$$G = \sqrt{(100)\ (200)} = \sqrt{20\ 000} \approx 141$$

which is somewhat smaller than the arithmetic mean of 150.

The harmonic mean

The harmonic mean is used to compute the mean of a series when the observations relate to rates of movement or growth. Using H to denote the harmonic mean, then

$$H = \frac{N}{\sum_{i} \frac{1}{X_i}}$$

3.8

For instance, if a truck delivering groceries from a warehouse to a supermarket averaged 20 mph on the outward trip, and 32 mph on the return trip over the same route, then the average speed, H, is

$$H = \frac{2}{\frac{1}{20} + \frac{1}{32}} = \frac{320}{13} = 24 \cdot 6 \text{ mph}$$

Measures of dispersion

Whereas measures of central tendency indicate where the *average* of a set of observations is to be found, measures of dispersion reflect the *spread* of a set of observations. In some cases, values are tightly clustered around the average position, while in other cases they are dispersed over a wide range. The four measures that will be reviewed describe the spread of a set of observations along a linear scale. As in the case of central tendency, there are also two-dimensional or bivariate measures of dispersion, the latter being of particular interest to centrographers. Only two of the more useful two-dimensional measures of dispersion will be discussed namely the *inter-neighbour interval* and the *standard distance:* other more esoteric measures may be found in Bachi (1957) and Neft (1966, Chapter 5).

The range

The range of a set of observations is the difference between the largest and smallest value, and hence is a measure of dispersion corresponding to the mode.

Suppose one has observations on a variable, Y, recording the yield of corn in bushels per acre in seven counties in the Mid West United States.

If the observations are

$$48.7 \quad 39.4 \quad 42.8 \quad 47.5 \quad 50.6 \quad 28.2 \quad 41.0$$

then the highest yield is 50·6, the lowest is 28·2, and the range is

$$50.6 - 28.2 = 22.4$$

Statisticians tend to be critical of the range as a measure of dispersion because of its sensitivity to extreme values. However, in certain circumstances it is the most appropriate measure of dispersion. Thus the range between high spring tide and low neap tide is the decisive measure in the construction of wharves and tidal docks, while the range between peak river discharge and minimum discharge is critical in the construction of dams and flood control systems.

The interquartile range

This ordinal measure of dispersion corresponds with the median. If the observations in a series are ranked from smallest to largest, then just as the median occupies the middle position, so the quartiles occupy the quarter positions, and the interquartile range is the difference between the values occupying the upper and lower quartile positions. If there are N observations in a series, then for the lower quartile (Q_1) we have (by logical extension of equation **3.1**)

$$\text{position of } Q_1 = \frac{N}{4} + \frac{1}{2} \qquad \textbf{3.9a}$$

while for the upper quartile (Q_3) we have

$$\text{position of } Q_3 = \frac{3N}{4} + \frac{1}{2} \qquad \textbf{3.9b}$$

In the case of variable Y recording corn yields, there are seven observations so that Q_1 and Q_3 occupy positions of 2¼ and 5¾ respectively. If we rank the values from smallest to largest we have

$$28.2 \quad 39.4 \quad 41.0 \quad 42.8 \quad 47.5 \quad 48.7 \quad 50.6$$

As the position of Q_1 is 2¼, it lies one-quarter of the distance between 39·4 and 41·0, so

$$Q_1 = 39.4 + \tfrac{1}{4}(41.0 - 39.4) = 39.8.$$

As the position of Q_3 is 5¾, this value lies ¾ of the distance between 47·5 and 48·7, so

$$Q_3 = 47 \cdot 5 + \tfrac{3}{4}(48 \cdot 7 - 47 \cdot 5) = 48 \cdot 4.$$

Finally, the interquartile range is given by

$$Q_R = Q_3 - Q_1 = 48 \cdot 4 - 39 \cdot 8 = 8 \cdot 6$$

It is worth noting that the interquartile range of Y is much less than half of the range (22·4).

The quartile is one of an infinite number of *quantiles* that can be used to measure dispersion. For instance one can use a 10—90 percentile or a 20—80 percentile range. Quantiles are not particularly useful measures of dispersion for they do not have the practical value of the range, nor do they occupy a position of prominence in statistical theory.

The mean deviation

The mean deviation $(\bar{D})$ is the arithmetic mean of the absolute deviations of a set of observations about their mean. Thus

$$\bar{D} = \frac{\Sigma \, |X_i - \bar{X}|}{N} = \frac{\Sigma \, |x_i|}{N} \qquad \textbf{3.10}$$

For the corn yields recorded in variable Y, the mean, $\bar{Y} = 42 \cdot 6$ and

$$\bar{D} = \frac{6 \cdot 1 + 3 \cdot 2 + 0 \cdot 2 + 4 \cdot 9 + 8 \cdot 0 + 14 \cdot 4 + 1 \cdot 6}{7} = \frac{38 \cdot 4}{7} = 5 \cdot 49$$

The mean deviation is rarely used as a measure of dispersion. As will become apparent shortly, the most appropriate measure of dispersion from a statistical viewpoint is the square root of the second moment about the mean, otherwise known as the standard deviation.

Standard deviation and variance

Equation **3·6** defines the first moment of a distribution. By applying the same formula, but raising the deviation scores to the power of two, one obtains the second moment of a distribution, which is more commonly known as the variance (s^2). Thus

$$s^2 = \frac{\Sigma \, (X_i - X)^2}{N} = \frac{\Sigma \, x^2}{N} \qquad \textbf{3.11}$$

Whereas the first moment of a distribution equals zero, the second moment involves squaring the deviation scores so that the variance is always a positive value.

The variance is a measure of great importance in parametric statistics; indeed much of this branch of statistics is concerned with accounting for variance in one form or another. However the most commonly used measure of statistical dispersion is not the variance, but the square root of the variance known either as the standard deviation (s), or as the standard error. From **3.11** we have

$$s = \sqrt{\frac{\Sigma (X_i - \bar{X})^2}{N}} = \sqrt{\frac{\Sigma x^2}{N}} \qquad \textbf{3.12}$$

Taking the example of corn yields recorded in variable Y, the standard deviation is

$$s = \sqrt{\left(\frac{6 \cdot 1^2 - 3 \cdot 2^2 + 0 \cdot 2^2 + 4 \cdot 9^2 + 8 \cdot 0^2 - 14 \cdot 4^2 - 1 \cdot 6^2}{7}\right)}$$

$$= \sqrt{\frac{345 \cdot 42}{7}} = \sqrt{49 \cdot 346} = 7 \cdot 02$$

The variance of Y is the square of the standard deviation, hence $s^2 = 49 \cdot 35$. It may be observed that equation **3.12** requires many subtractions: an alternative computational formula is presented in **3.21** below.

Measures of dispersion for two-dimensional distributions

From the various measures of spatial dispersion that are available, two are selected for discussion, the first because it can be given a simple physical interpretation, the second because it is closely related to the bivariate standard deviation.

The Inter-Neighbour Interval (INI), which has been devised by Court (1966), is simply the average distance between neighbours in a given area. If one wishes to express this average distance in feet, then

$$\text{INI} = 5280 \sqrt{A/N} \qquad \textbf{3.13}$$

where N is the number of persons in an area and
   A is the area expressed in square miles.

Two simple examples illustrate this measure: Gibraltar, with an estimated population in 1975 of 28 000 in an area of $2 \cdot 5$ square miles has an *INI* of 50 feet, whereas the French island of St Pierre to the south of Newfoundland has a population of 4800 in 10 square miles and the *INI* = 241 feet.

The measure of spatial dispersion that is most commonly used is known as the *standard distance* (Bachi, 1957). Using s_D to represent this parameter, then

$$s_D = \sqrt{s_x^2 + s_y^2}$$ 3.14

where s_x^2 and s_y^2 are respectively the variances of the northing and easting coordinates of the sample point pattern. It can be shown that the standard distance is the bivariate standard deviation multiplied by $\sqrt{2}$.

Higher-order moments

In discussing central tendency and dispersion, two moments have already been encountered, namely the first and second moments about the mean. The general formula for the rth moment is

$$\pi_r = \frac{\Sigma (X_i - \bar{X})^r}{N} = \frac{\Sigma x_i^r}{N}$$ 3.15

It is easy to see that when $r = 1$ one obtains **3.6** and when $r = 2$ one has the variance as defined in **3.11**. Two higher moments are commonly used, the third moment (π_3 – pi three) which is a measure of *skewness*, and the fourth moment (π_4 – pi four) which is a measure of *kurtosis* or peakedness.

Skewness indicates whether a distribution is symmetrical or lopsided. The simplest way to demonstrate these patterns is with a diagram. In Figure 3.3, A is negatively skewed, B is symmetrical, and C is positively skewed. These relationships are directly related to the values of π_3, so that a negative value indicates negative skewness, a positive value indicates positive skewness, and for a perfectly symmetrical distribution π_3 equals zero. Also shown in Figure 3.3 are the approximate locations of the mode, median, and mean for distributions with various kinds of skew. As can be seen, these measures of central tendency coincide in the case of a symmetrical distribution, but they diverge when skewness is present. It is instructive to calculate the third moment for variable Y (recording corn yields). We have

$$\pi_3 = \frac{\Sigma (X_i - \bar{X})^3}{N} = \frac{\Sigma x_i^3}{N}$$ 3.16

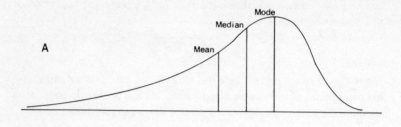

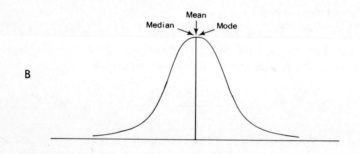

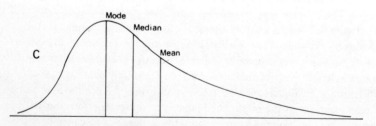

Figure 3.3: Skewed and symmetrical distributions and the location of the mode, median and mean.
A: Negatively skewed distribution (π_3 is negative).
B: Symmetrical distribution (π_3 is zero).
C: Positively skewed distribution (π_3 is positive).

hence

$$\pi_3 = \frac{6 \cdot 1^3 - 3 \cdot 2^3 + 0 \cdot 2^3 + 4 \cdot 9^3 + 8 \cdot 0^3 - 14 \cdot 4^3 - 1 \cdot 6^3}{7}$$

$$= \frac{-2166 \cdot 210}{7} = -309 \cdot 458$$

The third moment for Y is a negative value, therefore the distribution is negatively skewed as in Figure 3.3A. Furthermore we can check the consistency of this diagram because the mean of Y ought to be a smaller value than the median. $\bar{Y} = 42 \cdot 6$, whereas the median is $42 \cdot 8$: the expected relationship does indeed hold true.

The main weakness of π_3 is that it is an absolute quantity, so that with large numbers it can have a very large value. In order to convert π_3 to a relative value, one divides the third moment squared by the second moment cubed to obtain an index known as β_1 (beta one). Notationally,

$$\beta_1 = \frac{\pi_3^2}{\pi_2^3} \qquad \qquad \textbf{3.17}$$

In the present example

$$\beta_1 = \frac{-309 \cdot 458^2}{49 \cdot 35^3} = \cdot 7968$$

We will return to the β_1 index in the next chapter when tests of normality are discussed.

The fourth moment, π_4, is a measure of kurtosis. As Figure 3.4 illustrates, a flat curve is *platykurtic*, a normal curve is *mesokurtic*, while a peaked curve is *leptokurtic*. By applying **3.15** we have

$$\pi_4 = \frac{\Sigma (X_i - \bar{X})^4}{N} = \frac{\Sigma x_i^4}{N} \qquad \qquad \textbf{3.18}$$

which for the corn yield example gives us

$$\pi_4 = \frac{6 \cdot 1^4 - 3 \cdot 2^4 + 0 \cdot 2^4 + 4 \cdot 9^4 + 8 \cdot 0^4 - 14 \cdot 4^4 - 1 \cdot 6^4}{7}$$

$$= \frac{49166 \cdot 65}{7} = 7023 \cdot 81$$

As this example shows, π_4 can be very large because it is an absolute value. A relative index of kurtosis involves dividing π_4 by the variance

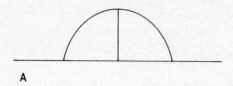

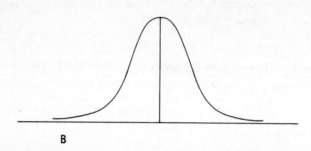

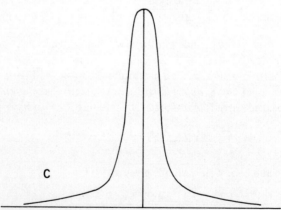

Figure 3.4: Types of kurtosis.
A: A platykurtic curve.
B: A mesokurtic curve.
C: A leptokurtic curve.

squared to obtain an index known as β_2 (beta two), i.e.,

$$\beta_2 = \frac{\pi_4}{\pi_2^2} \qquad\qquad \textbf{3.19}$$

and in the example

$$\beta_2 = \frac{7023 \cdot 81}{49 \cdot 35^2} = 2 \cdot 88$$

The β_2 index will be discussed further in the next chapter when tests of normality are examined, but for the present we will define curves with β_2 greater than 3 as leptokurtic, and those with β_2 less than 3 as platykurtic. The distribution of corn yields is therefore slightly platykurtic.

Estimating population parameters

Early in this book a distinction was made between a *sample* and a *population*. Thus far, the measures of central tendency and of dispersion that have been discussed relate to samples. These same measures may also be used to estimate the parameters of the population from which a sample is drawn. A good estimate of a population parameter should meet two main criteria (there are other criteria, but these will not be considered here):

1. It should be *unbiased.* Any individual estimate of a population parameter may be higher or lower than the true value because, after all, it is only an estimate. However when a whole series of estimates are obtained by drawing further samples from the population, in the limit one obtains a sampling distribution for the parameter that is being estimated. The sampling distribution of an unbiased estimate has a mean value (or more properly, a *mathematical expectation*) that is equal to the true value of the population parameter. If, on the other hand, the mathematical expectation of the parameter tends to be systematically either higher or lower than the true value, then that estimate is biased.

2. It should be *efficient.* The most efficient estimate is the one whose sampling distribution has the smallest possible dispersion: the greater the dispersion, the less efficient is the estimate.

The two parameters most frequently estimated are the population mean and standard deviation. These are discussed below following the convention of using Greek letters for population parameters and Roman letters for sample parameters.

First, we will consider how to estimate the population mean. Both the sample median, M, and the mean, $\bar{X}$, give unbiased estimates of the population mean, μ (pronounced mu). However the sample mean is a more efficient estimate than is the sample median because the variance of the sampling distribution of the means is less than that of the medians. In short, $\bar{X}$ is an unbiased and efficient estimate of μ.

Estimating the population standard deviation and variance is not quite so straightforward because s, as defined in **3.12**, is a biased estimate of the population standard deviation, σ (pronounced sigma). The numerator of **3.12**, $\Sigma(X_i - \bar{X})^2$, is known as the sum of the squares of deviations from the mean, or, more simply, the *sum of the squares*. Associated with any sum of squares is a quantity known as the *degrees of freedom*. For this sum of the squares there are $(N-1)$ degrees of freedom: this is because a set of N numbers can only vary about its mean in $(N-1)$ ways. For instance, if the mean of three numbers is known to be 5·0, and two of the numbers are 3 and 4, then the third number has to be 8. When $(N-1)$ numbers are specified, the Nth is determined. Applying this principle to estimates of the population standard deviation, an unbiased estimate of σ is obtained by using $(N-1)$ in the denominator. If we define a new term $\hat{s}$ (called s hat) where

$$\hat{s} = \sqrt{\frac{\Sigma(X_i - \bar{X})^2}{N-1}} \qquad \textbf{3.20}$$

then $\hat{s}$ is an unbiased and efficient estimate of σ. While **3.20** is a very literal equation, the formula most commonly used with calculators and in computer programs obtains $\hat{s}$ from

$$\hat{s} = \sqrt{\frac{\Sigma X^2}{N-1} - \frac{(\Sigma X)^2}{N(N-1)}} \qquad \textbf{3.21}$$

In **3.21** ΣX^2 involves squaring first and then summing, whereas $(\Sigma X)^2$ involves summing first and then squaring the total.

The same principles apply when computing the population variance, so that the usual machine formula for computing $\hat{s}^2$ is

$$\hat{s}^2 = \frac{\Sigma X^2}{N-1} - \frac{(\Sigma X)^2}{N(N-1)} \qquad \textbf{3.22}$$

and $\hat{s}^2$ is an unbiased and efficient estimate of the population variance, σ^2.

4 The binomial, Poisson and normal distributions

This chapter is concerned with the three most basic statistical distributions: two of these, the binomial and Poisson distributions, are *discrete* distributions (meaning that they describe events that occur as integers) whereas the normal distribution is a *continuous* distribution.

All three distributions are closely related to one another, although they are used in differing circumstances. The binomial distribution describes situations where one is dealing with a population of infinite size and there are two possible outcomes such as the head and tail of a coin. It is usual to make P (the probability of a 'successful' outcome) correspond with the outcome in which one is interested, and to make Q (the probability of a 'failure') correspond with the other outcome. The binomial distribution applies to samples of events which are neither rare, nor numerous. When the number of events (N) is large and P is a rare event, the binomial distribution is very closely approximated by the Poisson distribution. The Poisson distribution is used to describe rare events because in this situation the binomial formula is very tedious to evaluate. On the other hand when N is large, and an event is not rare, then the binomial distribution is approximated fairly closely by the normal distribution: and the larger is N, the better is the approximation until, in the limit when the classes are infinitesimally small, the binomial distribution and the normal distribution become identical. The availability of tables giving the area under the normal curve facilitates computations for the normal distribution; in contrast, laborious computations are involved in evaluating the binomial formula for large N, although tables are available in a few textbooks. Hence it is generally convenient to use the normal curve in situations where it provides a good approximation to the binomial distribution.

Statisticians tend to avoid actually defining such terms as 'large' and 'small', which is not very helpful for the beginner. At the risk of playing where angels fear to tread, the following rough and ready guidelines are suggested:

1. The Poisson distribution provides a good approximation to the

binomial distribution when the total number of events (N) is greater than 50, and the probability of P is so rare that $NP \leqslant 5$.

2. The normal distribution provides a good approximation to the binomial distribution when N is sufficiently large that *both $NP \geqslant 5$ and $NQ \geqslant 5$.*

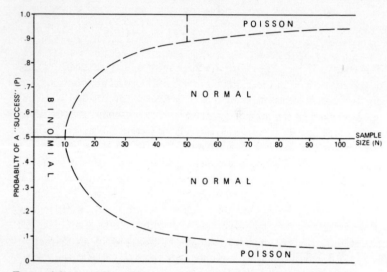

Figure 4.1: A guide to the use of the normal, binomial and Poisson distributions. The boundaries on the above diagram are only very approximate guides: different boundaries are recommended by different statisticians.

These relationships are plotted in Figure 4.1, in which the horizontal line shows sample sizes ranging from 0 to 100, and the vertical line probabilities from 0·0 to 1·0. Thus if $P = 0·8$, when $N = 40$ the normal distribution is recommended, but when $N = 20$ the binomial distribution is recommended, and so on.

The binomial distribution

The binomial distribution may be used for hypothesis testing, and to obtain point and interval estimates. The questions being posed in these three applications are, respectively:

1. Given the probability of a particular event occurring in a specified population, then what is the probability that a given sample was drawn from that known population?

For example, in throwing a fair dice, we expect a four to occur on average once in six throws: the probability of a four equals 0·1667, and the probability of all other numbers combined equals 0·8333. If, in throwing a given dice, a four occurs only once in thirty tosses, we may want to know what are the chances that the dice we are throwing is unfair.

2. Given the sample observations, what are the parameters of the population?
For instance, we may wish to estimate the variance of the population given sample data.

3. Given the probability of an event occurring, where are the confidence limits (for α level of significance) placed with a sample of size N?
For example, using the dice illustration again, in thirty tosses of the dice, where are the upper and lower confidence limits outside which we might suspect the dice to be unfair? If there were very few fours, or a large number of fours, in both cases we would consider the possibility of the dice being loaded.

So far, the binomial distribution has been treated as an *exact* distribution, that is, it exactly describes the situations being considered. In certain circumstances the binomial distribution may also be used as an approximation to the *hypergeometric* distribution. The hypergeometric distribution also describes situations where there are two possible outcomes, but unlike the binomial distribution, samples are drawn from a population of finite size. As a result, whereas the probabilities of success and failure are fixed in the case of the binomial distribution, in the case of the hypergeometric distribution these probabilities change as sampling occurs. Taking a simple example, suppose an urn contains 2 red and 2 green balls. Initially, the probability of drawing a green ball is 2 in 4, or ·5; but if one then drew out a green ball, the probability of drawing green at the next try is 1 in 3, or ·33.

When a population is large, the probabilities associated with various outcomes under the hypergeometric distribution are very similar to the corresponding probabilities for the binomial distribution; hence the latter distribution can provide a good approximation to the former. Insofar as the binomial distribution is easier to evaluate than the hypergeometric distribution, it is often advantageous to use the binomial approximation (the approximation results in a slightly conservative test).

Some arbitrary definition is needed of a 'good approximation'. The one adopted in this instance is a discrepancy in the size of the tails of the two distributions of not more than ½% when working at the ·05 level of significance. If the area in the tail of the hypergeometric distribution is 5% of the total area under the curve, the corresponding area for the binomial distribution must not exceed 5.5% of the total area. A computer program written by the author evaluating both the normal and hypergeometric distributions for different sizes of populations and samples revealed that, provided a sample is not larger than 6% of a population and the sample size (N) is greater than 4, the binomial distribution provides a good approximation to the hypergeometric distribution.

Before presenting the formula for the binomial distribution, the combinatorial expression $\binom{N}{X}$, which is sometimes written $_NC_X$, needs explaining. In Chapter 2 the sampling distribution for 3 tosses of a fair coin was evaluated. It was found, by tabulating every possible combination, that there are 3 ways that one tail can occur. We can use the combinatorial expression to find the same result because it tells us how many ways X events can occur in a sample of N events. The notation $\binom{N}{X}$ – (notice that this is *not* a division operation) – is shorthand for

$$\frac{N!}{X!\,(N-X)!}$$

To determine the number of combinations leading to two tails in four tosses of a coin, we evaluate

$$\frac{4!}{2!\,(4-2)!} = \frac{4!}{2!\,2!} = \frac{4 \times 3 \times 2 \times 1}{2 \times 1 \times 2 \times 1} = \frac{24}{4} = 6$$

When N is fairly large, it is worth taking advantage of the cancelling out properties of this expression. Suppose $N = 12$ and $X = 2$, then we have

$$\frac{N!}{X!\,(N-X)!} = \frac{12!}{2!\,(12-2)!} = \frac{12!}{2!\,10!}$$

but $12! = 12 \times 11 \times 10!$, hence by cancelling out we have

$$\frac{12 \times 11 \times 10!}{2!10!} = \frac{12 \times 11}{2} = 66$$

We are now in a position to introduce the formula for the binomial distribution (which is occasionally known as the Bernouilli distribution): we have

$$p(X) = \binom{N}{X} P^X Q^{N-X} \qquad\qquad 4.1$$

Where $p(X)$ = the probability of X outcomes in N events,
   P = the probability of a 'success',

Q = the probability of a 'failure'.

As $Q = (1 - P)$, it follows that $P + Q = 1.0$.

To illustrate this formula, suppose we wanted to know the probability of throwing two tails in five tosses of a fair coin. Then

$$p(2T) = \frac{5!}{2!\,(5-2)!}\ 0.5^2\ 0.5^{(5-2)} = \frac{5!}{2!\,3!}\ 0.5^2\ 0.5^3$$

$$= 10 \times .25 \times .125 = .3125$$

The hypergeometric distribution also makes use of this combinatorial expression. If

A = the number of 'successful' outcomes in the original population

B = the number of 'unsuccessful' outcomes in the original population

and $A + B = C$ = the total number of observations in the population

then

$$p(X) = \frac{\binom{A}{X}\binom{B}{N-X}}{\binom{C}{N}} = \frac{\left(\frac{A!}{X!\,(A-X)!}\right)\left(\frac{B!}{(N-X)!\,(B-N-X)!}\right)}{\frac{C!}{N!\,(C-N)!}} \qquad \textbf{4.2}$$

Clearly, this expression involves many more combinatorial terms than does the binomial formula, and it is for this reason that the binomial approximation is recommended in certain circumstances.

Two constants of the binomial distribution will now be introduced, namely the mean and the standard deviation (standard error):

the mean $= NP$ **4.3**

the standard deviation $= \sqrt{NPQ}$ **4.4**

These constants will be used to demonstrate the normal approximation to the binomial distribution in the following chapter.

The Poisson distribution

In contrast to the coin-tossing examples used in many statistics books to illustrate the binomial distribution, the most renowned application of the Poisson distribution is decidedly interesting. It was used by Count von Bortkiewicz to describe the number of cavalrymen in certain Prussian cavalry corps who died as a result of horse kicks between 1875 and 1894. Despite the bizarre nature of this application it does illustrate two aspects of the Poisson distribution: it is a discrete distribution used to describe events that occur as integers; and it is used to describe rare events. Moroney (1956) in his book *Facts from Figures* gives the title

'Goals, floods and horse kicks' to his chapter dealing with the Poisson
distribution, thereby indicating two other rare events fairly well
described by this distribution.

The Poisson distribution is used under the following conditions:
$N \geqslant 50$ and $NP \leqslant 5$. When $NP > 5$ it is common to use the normal
distribution with large samples and the binomial distribution with small
samples. The Poisson distribution may also be used when the number
of unsuccessful outcomes is unknown, and in this respect it differs from
the binomial distribution. In the case of the cavalry officers we know
how many were killed and might possibly find out how many were not
killed, but in the case of points in space there are an infinite number of
locations where a point is not present. In this situation we cannot use
the binomial distribution because the probability of a failure is indeter-
minate; a problem which does not arise with the Poisson distribution.

The formula for the Poisson distribution is given by

$$p(X) = \frac{\lambda^X e^{-\lambda}}{X!} \qquad \textbf{4.5}$$

where $p(X)$ is the probability of X events occurring in a given sample;
  P is the probability of an event occurring;
  N is the total sample size;
  $\lambda = NP =$ the mean and the variance of the Poisson distribution;
  $e =$ the mathematical constant which forms the base of the
  natural logarithms, calculated by:

$$e = \frac{1}{0!} + \frac{1}{1!} + \frac{1}{2!} + \frac{1}{3!} + \frac{1}{4!} + \ldots + \frac{1}{\infty!} = 2 \cdot 71828 \ldots$$

To illustrate this formula, suppose that the probability of an event
occurring is 1 in 100 times, and the sample we are dealing with consists
of 200 observations (i.e. $P = \cdot 01$, $N = 200$ and $\lambda = NP = 2$). The
probability of exactly one event occurring is

$$p(1) = \frac{2^1 e^{-2}}{1!} = \frac{2}{e^2} = \cdot 2707$$

Note how cumbersome the binomial formula would be: in the same
situation we have

$$p(1) = \frac{200!}{1! \; 199!} \cdot 01^1 \cdot 99^{199}$$

The term $e^{-\lambda}$ of the Poisson distribution can be difficult to evaluate, hence a table giving values for this term is generally used (see Table A, page 253.

There is a short-cut method for evaluating the Poisson distribution. Suppose one wishes to evaluate $p(2)$ and $p(3)$ and $\lambda = 0.9$. Then

$$p(2) = \frac{0.9^2 \, e^{-0.9}}{2!} = .1647$$

$$p(3) = \frac{0.9^3 \, e^{-0.9}}{3!} = .0494$$

But if we rewrite $p(3)$, expanding 0.9^3 to 0.9×0.9^2, and $3!$ to $3 \times 2!$, then

$$p(3) = \frac{0.9}{3} \times \frac{0.9^2 e^{-0.9}}{2!} = \frac{0.9}{3} \times p(2) = .0494$$

and in general

$$p(X) = p(X-1) \times \frac{\lambda}{X} \qquad \textbf{4.6}$$

The one drawback to this short cut is that any error made in the first calculation is carried through all the subsequent calculations. A useful check is to compare the end value of a chain to the probability expected using the long method: these values should be in very close agreement.

Deviation scores and standard scores

Before discussing the normal distribution, a brief explanation is needed of deviation scores and standard scores. Using upper-case letters to represent raw scores and lower-case letters to represent deviation scores, then

$$x_i = X_i - \bar{X} \qquad \textbf{4.7}$$

A deviation score indicates whether any individual observation is larger or smaller than the mean, and by what magnitude. Deviation scores are recorded in the original units of measurement; this makes for easy comparisons between two variables recorded in the same units over the same range of magnitude, but does present a problem when two different variables are recorded. Suppose one wished to compare household income with car ownership: the first variable might have a range from $100 to $100 000, while the second might range from $0 - 4$ cars. How can these two variables be converted into a common basis of measurement? The answer is to divide the deviation scores by the standard deviation to obtain *standard scores* (otherwise known as z scores).

Hence

$$z_i = \frac{(X_i - \bar{X})}{s} = \frac{x_i}{s} \qquad 4.8$$

The z scores for a sample or a population have two important properties:

1. Their mean equals zero.
2. Their standard deviation and variance equals one, or, more formally, they have *unit variance*.

By converting raw scores to standard scores two very dissimilar variables can be converted to a common basis of measurement. Not only do z scores have this practical value, but they are also important in a theoretical sense. The normal distribution, the appropriate sampling distribution for a number of statistical tests, is completely specified in terms of z values: so in appropriate cases one can use z values to make probabilistic statements.

The normal distribution

As Figure 4.1 (page 56) demonstrates, the normal or Gaussian distribution is closely related to the binomial and Poisson distributions. Several other distributions besides the binomial and Poisson are closely approximated by the normal curve. Since there are tables of the probabilities associated with specific z values under the normal curve, it is often quicker to use the normal approximation than to calculate an exact probability. Moreover, the normal distribution not only approximates a number of other distributions but also exactly describes the sampling distribution of several statistics. In short, the normal distribution can often be used as a 'model' to which empirical results are compared. It follows that the normal distribution is of great importance as a sampling distribution in inferential statistics.

The importance of the normal distribution in statistics rests not only on its theoretical value, but also derives from its empirical value. There are a whole host of phenomena with frequency distributions that are roughly normal (do not, however, make the mistake of assuming that all things in nature are normally distributed). Perhaps of equal importance to geographers is another set of phenomena that have positively skewed frequency distributions approximating the lognormal or Gibrat distribution; simply by taking the logarithm of each observation one can transform such a curve to a shape that is approximately normal.

In discussing standard scores it was stated that the normal curve is

completely specified in terms of z values. In its standard form, the curve
Y is given by

$$Y = \frac{1}{\sqrt{2\pi}} \, e^{-\frac{1}{2}z^2} \qquad\qquad \textbf{4.9}$$

where $\pi = 3 \cdot 1416 \ldots$.
and e (the base of the natural logarithms) $= 2 \cdot 71828 \ldots$.

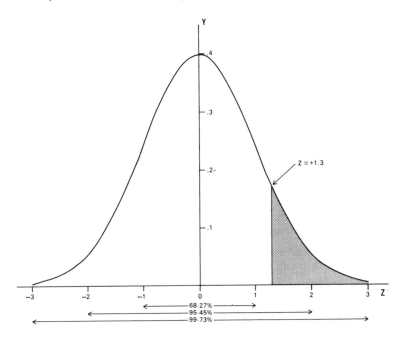

Figure 4.2: The normal curve.

The curve, which is plotted in Figure 4.2, has certain characteristics that
merit discussion. First, Y is conventionally plotted for z values from
−3 to +3: as z values become more extreme so Y asymptotically
approaches zero (that is, it approaches zero without ever quite reaching
zero). Second, the curve is symmetrical: exactly half of the area under
the curve lies on each side of zero. Third, the area under the curve is
known: for instance 68·27% of the area lies between $z = -1$ and $z = +1$;
95·45% of the area lies between $z = -2$ and $z = +2$; while 99·73% lies
between $z = -3$ and $z = +3$.

The exact area under the curve between any two values of z, say a
and b, is given in Table F (page 258). The total area under the curve is

unity, hence z values can be expressed as probabilities. For instance, the shaded area in Figure 4.2 lying between $z = +1\cdot3$ and $z = +\infty$ occupies 9·7% of the total area under the normal curve, hence a value $z \geqslant 1\cdot30$ has a probability of ·097 of occurring.

Testing for normality

How does one decide whether or not an empirical distribution is normally distributed? There are various methods for tackling this problem, ranging from simple checks to formal tests. Four such methods will be reviewed here using sample data recording length (in kilometres) of a journey to a shop for shoes. The data, which will be referred to as variable A, consist of the 25 observations in the table:

5·57	7·23	10·02	10·44	12·22
13·96	14·74	14·78	15·06	15·50
15·71	16·31	17·08	17·78	18·34
19·12	19·67	20·04	21·65	23·10
23·47	24·93	26·11	26·31	27·81

1 Histogram

Plot a histogram and inspect the shape of the resulting frequency distribution. This 'eyeball' method is very simple, yet once one has some experience with frequency distributions it provides a surprisingly good guide. Besides, the plotting of a histogram is recommended even if one intends to proceed to one of the more formal tests of normality. In order to plot a histogram, three steps should be followed.

Step 1. Choose the number of classes, c, where c is approximately the square root of the number of observations. Thus with 25 observations, 5 classes would be used.

Step 2. Identify the range of the observations, pick some arbitrary but convenient lower and upper limits which lie slightly outside the range, and divide the new arbitrarily defined range into c equally sized classes. In variable A the observations range from 5·57 to 27·81 and there are five classes, hence it is convenient to define arbitrary lower and upper limits of 5·50 and 28·0, respectively, and to use a class interval of 4·5 so that the class boundaries have values of 5·5, 10·0, 14·5, 19·0, 23·5 and 28·0.

Step 3. Count the number of observations that fall in each class interval and plot a histogram. Figure 4.3 is the histogram for variable A, and it can be seen that the data are roughly normal, although some slight negative skew is present.

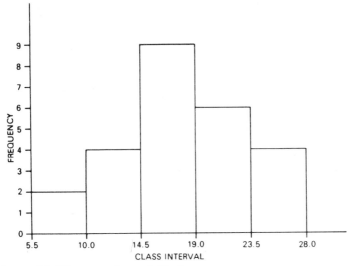

Figure 4.3: Histogram for variable A.

2 Cumulative curve

Plot the cumulative frequency distribution on a special type of graph paper known as *arithmetic probability paper*. Figure 4.4 illustrates how this is done. The vertical scale records the cumulative percentage of observations on a variable scale so that, for instance, the vertical distance from 50% to 60% is similar to the vertical distance from 1% to 2%. Note that the vertical scale is drawn from 0·5% to 99·5%: the limiting values of 0% and 100% are infinitely distant. The horizontal scale is plotted in regular units for the range of the variable concerned. Thus the cumulative distribution in Figure 4.4 shows that 50% of the shopping trips were of 17 kilometres or less. The constant horizontal and variable vertical scale has the effect of turning the curves of a cumulated normal distribution into a straight line.

3 Chi square test

The chi square test may be used as goodness-of-fit test for comparing a sample to the normal distribution: examples may be found in Croxton, Cowden and Klein (1968, pp. 595–6) and Yeates (1974, pp. 202–3). There is one important problem with this test: in order to meet the criteria for expected frequencies that will be outlined in Chapter 6, one often has to amalgamate the categories occupying the tails of a dist-

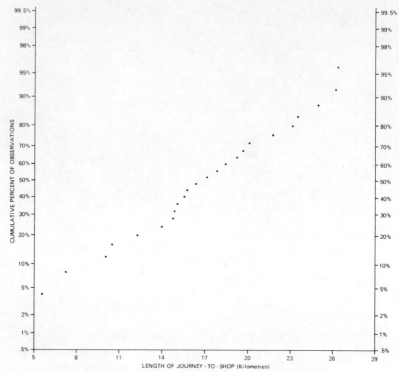

Figure 4.4: Cumulative curve for variable A *using arithmetic probability paper.*

ribution. Unfortunately, it is precisely in the tails that skewness is most likely to be present, hence the test tends to be somewhat insensitive. If this test is used, it is important to remember that the sample is made to correspond with the normal distribution with respect to N, $\overline{X}$ and s, hence three degrees of freedom are lost, one for each parameter estimated from the sample.

4 Formal test of skewness and kurtosis

In theory, the best method to test for normality is to make a formal test on the amount of skewness and kurtosis present. In the previous chapter two standardized measures were presented: β_1, defined in **3.17**, is a standardized measure of skewness; and β_2, defined in **3.19**, is a standardized measure of kurtosis. In the case of a normal distribution, $\beta_1 = 0$ and $\beta_2 \approx 3$. Egon Pearson has tabulated the confidence limits for these two measures, and a modified version of Pearson's table may be found in Table G, page 259. In the version of this table presented here, critical values have been computed for certain values of N smaller than those given by Pearson, while values for very large N have been omitted.

The main weakness of this test is that it is not reliable when the sample size is small. Take the β_1 coefficient measuring skewness and suppose $N = 50$: if β_1 falls outside the critical limits then one can assume that the distribution is non-normal, but if β_1 falls within the critical limits the distribution is not necessarily normal since several other distributions give rise to an apparently acceptable β_1 value. β_2 is even more difficult to interpret: the sampling distribution of β_1 converges on the normal form quite quickly and is approximately normal for $N \geqslant 100$, but the sampling distribution of β_2 even for $N = 1000$ is somewhat skewed and leptokurtic. In short, although Pearson's test provides the necessary and sufficient conditions for normality when N is very large, it provides only the necessary conditions, in the case of β_1, when $N < 100$, and in the case of β_2 when $N < 1000$.

Applying Pearson's test to variable A: $\beta_1 = \cdot0083$ and $\beta_2 = 2\cdot382$. Both values lie well within the critical limits. Such a result is consistent with a normal distribution, but since $N = 25$, this does not guarantee that the data are indeed normal.

Normalizing data

Having plotted a histogram, and perhaps having conducted a formal test for normality, what does one do if the data are clearly non-normal? The standard answer is to transform data, but this raises a contentious issue. On the one hand there are statisticians who argue that transforming data to normalize the frequency curve is nothing more than 'fudging' the data to fit the model, and that the implications of transforming data are not fully understood. On the other hand there are statisticians who argue that all measurement systems are arbitrary: hence transformed data are just as valid as untransformed data. This latter group have no hesitation in using a transformation to normalize data if normally distributed data are required.

Many of the frequency distributions used by geographers are known to be skewed with an attenuated upper tail. For instance, city, plant and firm size and various forms of interaction over distance tend to have approximately lognormal frequency distributions. If transformation of these data to a normal form were not valid, then geographers would be faced with almost intractable problems in applying parametric methods using these types of data. In consequence geographers have generally made extensive use of data transformations. While that in no sense legitimizes the practice, the discipline would have to discard much of the analytical research accumulated over the last fifteen years if the use of transformations were demonstrated to be invalid.

For positively skewed data one needs a transformation that will reduce values in the upper tail by a greater amount than those located in the lower tail. A fairly moderate transformation is achieved by taking the square root of each observation; this is often done with data involving distances and areas. A more radical transformation is achieved by taking the logarithm of each observation. Suppose we have three observations: $X_1 = 1 \cdot 26$, $X_2 = 2 \cdot 51$, and $X_3 = 5 \cdot 01$. The range from X_1 to X_2 is $0 \cdot 75$, and from X_2 to X_3 is $2 \cdot 50$. However, taking common logarithms the observations become $0 \cdot 1$, $0 \cdot 4$ and $0 \cdot 7$, respectively, and the range separating the observations becomes identical. In cases of extreme skewness, double logarithmic transformations may sometimes be applied. Logarithmic transformations cannot be applied to an array containing one or more negative values; before the logarithms are obtained a constant has to be added to all of the observations to eliminate negative values.

Much less frequently, negatively skewed data are encountered; a logical extension of the above discussion shows these may be transformed by squaring values or taking anti-logarithms. Difficulties arise, however, when data have unusually shaped frequency distributions. For instance, King (1961) has discovered that towns tend either to specialize or not to specialize in manufacturing, so that the distribution describing the proportion of employees engaged in manufacturing activity in a system of cities tends to be U-shaped, often with an arc-sine distribution. No amount of logarithmic or square root transformations will normalize this distribution. To be sure, one can normalize any distribution by concocting an elaborate transformation function, but — outside of the basic transformations outlined above and possibly the probit and logit transformations discussed by Bartlett (1947) — diminishing marginal returns to effort rapidly set in. Where, then, does one draw the line? The answer is not easy to prescribe, but if one can convert a non-normal distribution to an approximately normal form without a complicated transformation, the effort is probably warranted. Indeed many parametric methods, including those based on the F distribution, work fairly efficiently with rectangular frequency distributions; so, provided skewness can be removed, a transformation may be warranted. If, in contrast, it is difficult to remove skewness or other irregularities, then it is best to fall back on an appropriate nonparametric test.

Deviant observations

When examining frequency distributions one quite commonly encounters deviant observations which fall well outside the range for the remainder of the observations. The treatment of these 'outliers' is difficult because they can be attributed to three quite different causes:

1. They can result from measurement or human error: these errors include such things as mispunched data cards, faulty measurement devices and careless arithmetic.

2. They can be attributed to specification error: this gives rise to a *bona fide* extreme case that occurs for some specifiable reason. Suppose for instance, one were examining the influence of ice action on the size of cirques and one cirque was found to be much larger than the others because it was developed in a major structural weakness. This deviant observation is explicable in terms of a variable that is not part of the specified ice-action mechanism.

3. They can be true random extreme values.

In the first two cases it is legitimate to drop deviant observations from a sample, but in the third case it is incorrect. Unfortunately it is extraordinarily difficult to lay down ground rules for distinguishing the three causes of deviant observations. At the same time it is very important to distinguish them because the moments of a distribution are strongly influenced by extreme values. The best that can be suggested is to trace the data back to its original source, apply common sense and, where appropriate, consult *The Guinness Book of Records.*

Part two
Nonparametric methods

5 One sample nonparametric tests using the binomial and Poisson distributions

The binomial and Poisson distributions describe the probabilities of various outcomes under certain defined assumptions. Accordingly, they serve as sampling distributions to which empirical data can be compared, and hence give rise to two simple statistical tests. Both are one sample tests because a single sample is compared to the chosen theoretical distribution: they are nonparametric because no assumptions are made about the distribution of the parent population.

The binomial test

Statistical applications

The binomial test may be used as an exact test or as an approximation to situations described by the related hypergeometric distribution (in the previous chapter the approximation was deemed 'good' if the sample exceeds $4(N > 4)$ and is not larger than 6% of the population). In either case, the test described situations in which there are two outcomes: heads or tails, urban or rural, wet or dry, and so on. One needs to know the probabilities of the two outcomes, P and Q, in the population as a whole. One can then assess the probability that a sample consisting of an observed number of dry days, or whatever, was drawn from that specified population. The binomial test is normally used when either NP or $NQ < 5$, and when neither P nor Q are rare events.

Assumptions (under the null hypothesis)

Data should represent discrete events measured at the nominal or ordinal scale. It is possible to dichotomize interval scale data so as to use the binomial test, but this generally involves a loss of information.

Exact test: observations are drawn from an infinite population and should be independent so that the chances of P and Q are the same for each event, and the chances of P and Q remain constant throughout the sampling.

Approximate test: observations are drawn from a finite population. At any stage in the sampling each individual that has not been drawn has the same chance of being drawn.

Approximate test: the number of observations in a sample should be greater than 4, and should not exceed 6% of the population if the approximation is to be 'good'.

Test statistic

In order to conduct the binomial test one has to evaluate the binomial distribution as given by equation **4.1** in the previous chapter, where

$$p(X) = \binom{N}{X} P^X Q^{N-X} \qquad \textbf{4.1}$$

Initially it is instructive to evaluate the whole distribution, but with experience, it is necessary only to evaluate the tails (or single tail in the case of a one-tailed test).

Having evaluated the sampling distribution one is then in a position to perform the computations leading up to rejection or acceptance of the null hypothesis. These computations are done in the following stages:

1. Determine the exact location of the critical regions as follows. If the test is two-tailed, α is equally divided between the two tails. If the test is one-tailed, the research hypothesis (H_1) will indicate in which tail the critical region is located: if an above average frequency is hypothesized, it is located in the upper tail, while a below average frequency involves the lower tail.

2. In a one-tailed test, identify the extreme outcome in the tail associated with the critical region, and sum together the probabilities beginning with the extreme outcome and ending with the observed event. For instance, if 11 events occurred in a sample of 13 observations and the critical region was in the upper tail, one would add together $p(13) + p(12) + p(11)$. If the result is less than α, the observed outcome lies *wholly within* the critical region and H_0 is rejected.

3. In the case of a two-tailed test, ascertain which tail the observed outcome is closest to, and then proceed to sum together the probabilities in that tail, beginning with the extreme outcome and ending with the observed event. If the result is less than $\alpha/2$, reject H_0.

Example

The example that is developed here makes use of the binomial distribution as an approximation to the hypergeometric distribution. The main points of the problem are summarized in Figure 5.1. We are dealing with house ownership in an area in which there are 250 properties. Recorded on the map are changes in house ownership between the time when the lots were first sold in 1959 and the time of data collection in May 1972. Although the lots were not all sold simultaneously in 1959, they were sold over a fairly short period so that for all intents and purposes we are examining changes in ownership over a thirteen-year period. Two outcomes are considered: a house may have its original owner (a 'success'); or it may have undergone at least one change of ownership (a 'failure'). It transpires that 130 houses have their original owner, and 120 houses do not, so that in the population as a whole, $P = \cdot52$ and $Q = \cdot48$.

Within this population, samples can be drawn with respect to a variety of characteristics; for example, asterisks indicate corner properties, while the letters N, S, E and W indicate the cardinal orientation of the houses. A number of construction firms were involved in building in this area: the sample selected in this example is composed of all eight houses built by a firm reputed to have done an inferior job so that these houses have damp basements and associated problems. These eight houses are located on the north side of Weller Crescent (excluding the corner lot) and are identified on Figure 5.1 as Sample A. Seven out of the eight houses have undergone at least one change of ownership. The question that is posed is whether this rate of turnover could have occurred by chance: we suspect not, because of the rumoured faulty job of construction.

PROBLEM DEFINITION

1. H_0: the proportion of original house owners in the sample is not smaller than the proportion in the population as a whole.

H_1: there is a smaller than average proportion of the original house owners in the sample.

The test is one-tailed.

2. The level of significance, $\alpha = \cdot05$.

TEST SELECTION

1. There are two categories, and measurement is at the ordinal level.
2. The sample size, $N = 8$.

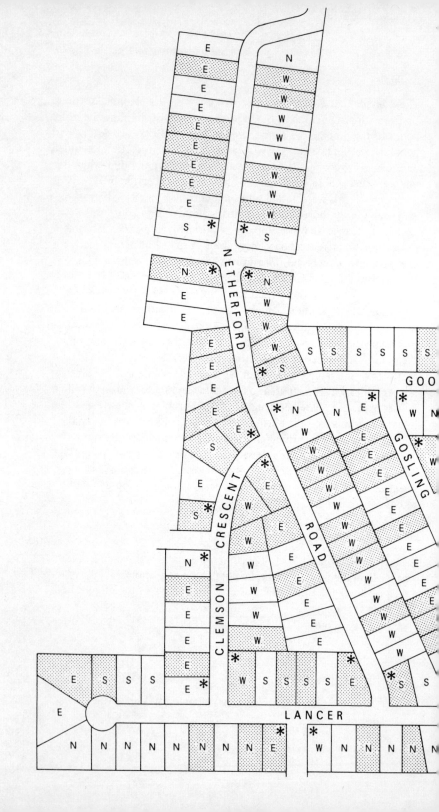

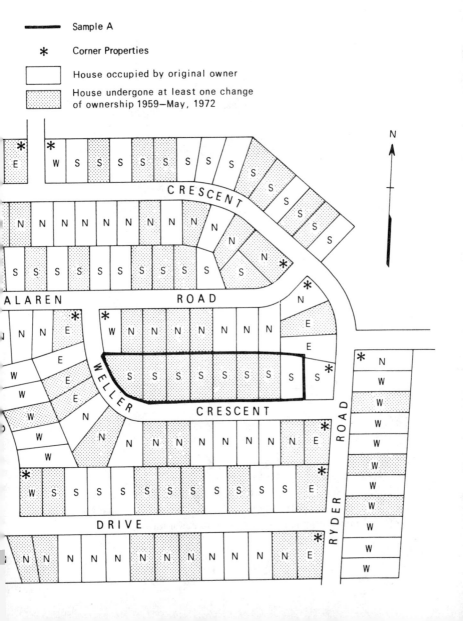

Figure 5.1: Changes in house ownership in a subdivision during a thirteen-year period.

3. For a sample of 8 drawn from a population of 250, the binomial distribution gives a good approximation to the hypergeometric distribution: indeed the difference between these two distributions in this case is so small that it is difficult to distinguish one from another on a graph (see Figure 5.2).

SAMPLING DISTRIBUTION

Given a sample of 8 houses, then there are 9 possible outcomes, from $X = 0$ through to $X = 8$.

Evaluating the binomial distribution with $N = 8, P = \cdot52, Q = \cdot48$ and $X = 0, 1, 2, \ldots, 8$, and bearing in mind the definitions that $0! = 1$ and $X^0 = 1$ then we have

$$p(0) = \frac{8!}{0!\,8!} \;\cdot52^0 \;\cdot48^8 = 1 \times 1 \times \cdot0028 = \qquad \cdot0028$$

$$p(1) = \frac{8!}{1!\,7!} \;\cdot52^1 \;\cdot48^7 = 8 \times \cdot52 \times \cdot0059 = \qquad \cdot0245$$

$$p(2) = \frac{8!}{2!\,6!} \;\cdot52^2 \;\cdot48^6 = 28 \times \cdot2704 \times \cdot0122 = \quad \cdot0924$$

$$p(3) = \frac{8!}{3!\,5!} \;\cdot52^3 \;\cdot48^5 = 56 \times \cdot1406 \times \cdot0255 = \quad \cdot2008$$

$$p(4) = \frac{8!}{4!\,4!} \;\cdot52^4 \;\cdot48^4 = 70 \times \cdot0731 \times \cdot0531 = \quad \cdot2717$$

$$p(5) = \frac{8!}{5!\,3!} \;\cdot52^5 \;\cdot48^3 = 56 \times \cdot0380 \times \cdot1106 = \quad \cdot2354$$

$$p(6) = \frac{8!}{6!\,2!} \;\cdot52^6 \;\cdot48^2 = 28 \times \cdot0198 \times \cdot2304 = \quad \cdot1277$$

$$p(7) = \frac{8!}{7!\,1!} \;\cdot52^7 \;\cdot48^1 = 8 \times \cdot0103 \times \cdot48 = \qquad \cdot0396$$

$$p(8) = \frac{8!}{8!\,0!} \;\cdot52^8 \;\cdot48^0 = 1 \times \cdot0053 \times 1 = \qquad \cdot0053$$

These results are plotted in Figure 5.2.

We are now in a position to check the calculations: the total area under a sampling distribution equals one, hence the probability of all the outcomes summed together should equal unity. Here

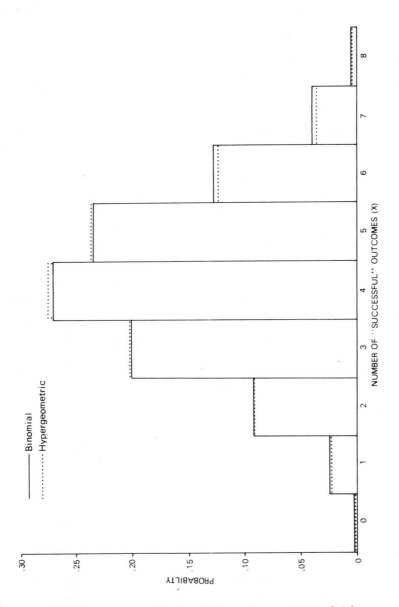

Figure 5.2: Sampling distribution of all samples consisting of eight observations for the binomial distribution when P = ·52 and Q = ·48, and the hypergeometric distribution when A = 130 and B = 120.

$$\sum_{X=0}^{8} p(X) = 1\cdot0002$$

which is correct, subject to a small rounding error.

COMPUTATION

In order to reject the null hypothesis we need the sample outcome (one out of eight houses still belonging to the original owner) to lie wholly within the critical region. To do this, we begin with the extreme out-come in the tail and sum together the probabilities until the observed outcome is reached. It happens that the second outcome is the observed one; hence we need sum only two probabilities. If the sum of these probabilities is less than the critical value, we reject the null hypothesis because the observed outcome lies wholly within the critical region. More formally, in the present case

$$\text{if } p(0) + p(1) < \alpha \text{ we reject } H_0.$$

$$\text{As } \cdot0028 + \cdot0245 = \cdot0273$$

we reject the null hypothesis, and conclude that there is a significantly smaller than average proportion of the original owners in this sample.

Geographical applications

King (1969, p. 40) states:

the binomial law has not proved useful in geographic research. The fact that for most spatial patterns the probability associated with any locational event usually is fairly small means that another discrete distribution, the Poisson, is more appropriate.

Taking published geographical literature as a guide, King's statement must be accepted. When the probability associated with an event is small, such as in point pattern analysis, the Poisson distribution has been commonly used, and with large samples, the normal approximation to the binomial distribution has been applied. Furthermore, in a variety of geographical situations sampling takes place from a finite population, in violation of one assumption of the exact binomial test.

There are two reasons for suggesting that King's assessment may be somewhat harsh, and for upgrading the utility of the binomial test, while still accepting that the range of applications is fairly limited. First, geographers have a predilection for working at the macro-scale with large samples, to the extent that micro-scale studies and the associated small samples are underrepresented. There would appear to be a shift towards micro studies in such fields as social geography, and this may result in new uses for the binomial distribution being identified in the future. Second, the binomial approximation to the hypergeometric distribution opens up a variety of geographical applications that have hitherto been neglected.

The binomial distribution is a special case of the *multinomial distribution.* In the binomial case there are two possible outcomes, in the multinomial there are more than two possible outcomes. The multinomial distribution has been rarely used by geographers and will not be discussed further here.

An example of the normal approximation to the binomial distribution

In this example, the data used to illustrate the binomial test will be re-worked, using the normal distribution as a theoretical model to which the data are compared. The formal procedure for conducting a test will not be used as it has already been spelled out. The example involved a sample of eight houses, only one of which was still owned by the original purchaser, yet in the population as a whole, the probability of a house being owned by the original purchaser $(P) = \cdot52$, and the probability of at least one change of ownership taking place $(Q) = \cdot48$.

Figure 5.3 shows the binomial distribution for this example, with a normal curve superimposed on the discrete distribution. The reason for presenting this diagram is to illustrate why a *correction for continuity* is applied when using the normal approximation to the binomial. The probability used in the original binomial test was the area represented in the bars above 0 and 1. Now in the normal approximation, if the area under the normal curve to the left of $X = 1$ were used, the probability would be underestimated. Quite clearly, the required area under the normal curve lies to the left of the point where $X = 1\cdot5$ because this is the mid-point between the two discrete outcomes, $X = 1$ and $X = 2$.

Equation **4.8** states that a z value may be obtained from

$$z_i = \frac{X_i - \overline{X}}{s} \qquad\qquad 4.8$$

In the present example, after correcting for continuity, $X_i = 1\cdot5$. The two parameters, $\overline{X}$ and s were defined for the binomial distribution in Chapter 4:

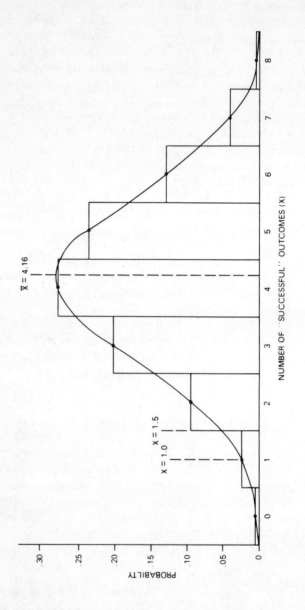

Figure 5.3: Normal approximation to a binomial distribution.

$$\text{the mean} = \bar{X} = NP$$

$$\text{the standard deviation} = s = \sqrt{NPQ}$$

Evaluating these parameters, we have

$$NP = 8 \times \cdot 52 = 4 \cdot 16$$

$$\sqrt{NPQ} = \sqrt{8 \times \cdot 52 \times \cdot 48} = \sqrt{1 \cdot 9968} = 1 \cdot 413$$

With this information, z is given by

$$z = \frac{1 \cdot 5 - 4 \cdot 16}{1 \cdot 413} = \frac{-2 \cdot 66}{1 \cdot 413} = -1 \cdot 88$$

Consulting Table F in the Appendix (page 258), the probability associated with $z = -1 \cdot 88$ in a one tailed test is $\cdot 0301$. It is worth stressing how close this is to the exact probability of $\cdot 0273$, computed according to the binomial formula, even though the sample consists of only eight observations (which is less than the minimum recommended on page 56).

The Poisson test

Statistical applications

The Poisson test is used to determine whether a sample consisting of rare events (in which $N > 50$ and $NP \leqslant 5$) could have been drawn from a population in which the probability of the rare event is known.

Assumptions

The observations should be discrete events measured at the nominal or ordinal scale.

Events should be independent so that

1. the occurrence of one event in a sample must not affect the probability of another event occurring in that sample or in any other sample;

2. each sample has the same per caput or per unit chance of an event occurring in it;

3. no particular event is more strongly attracted to one sample than to any other sample.

Test statistic

The Poisson test involves evaluating the formula for the Poisson distribution, given in **4.5** as

$$p(X) = \frac{\lambda^X e^{-\lambda}}{X!} \qquad \qquad 4.5$$

However, in conducting a Poisson test, one does not evaluate the whole sampling distribution. Using Y to represent the observed number of outcomes in a sample of size N, then the part of the sampling distribution that is needed is determined by the following rules.

First, the somewhat simpler case of a one-tailed test is considered. H_1 will indicate whether the critical region lies in the upper or lower tail. If the lower tail is involved one simply sums together the probabilities from $p(0)$ to $p(Y)$ inclusive, and if the result is less than α, H_0 is rejected. If the upper tail is involved, one sums together the probabilities from $p(0)$ to $p(Y-1)$ and subtracts this value from 1: if the result is less than α, H_0 is rejected. Notice, in the case of the upper tail of the binomial distribution, that one begins with the extreme event in the upper tail and sums together the probabilities down as far as $p(Y)$. This is clearly an inefficient method in the case of the Poisson distribution: for instance if $Y = 5$ and $N = 100$, it is far easier to sum from $p(0)$ to $p(4)$ inclusive and subtract the result from one than it is to sum all the probabilities from $p(5)$ to $p(100)$ inclusive.

In the case of a two-tailed test, the mean, λ, is needed. If Y is less than the mean (i.e. $Y < \lambda$) then one works with the lower tail, while if $Y > \lambda$ one works with the upper tail. In both cases the corresponding procedure for evaluating a one-tailed test is followed, except that the summed probabilities are compared to $\alpha/2$.

Example

The example that is used is drawn from Mieczyslaw Choynowski's famous study of the incidence of brain tumours in Poland. In the sixty *poviat* (counties) that he studied in Southern Poland, the average incidence was 5·17 per 100 000 inhabitants. Clearly, this is a rare event ($P = \cdot0000517$) and the disease occurs as discrete cases. The problem which Choynowski explored was whether the incidence of brain tumours in any of the *poviat* was significantly higher or lower than in the population as a whole. For the purposes of this example, three *poviat* (Brzozow, Gorlice and Lesko) will be examined.

PROBLEM DEFINITION

1. H_0: the incidence of brain tumours in each *poviat* does not differ significantly from the frequency in the population as a whole.

H_1: the incidence of brain tumours in certain *poviat* is significantly different from the incidence in the population as a whole.

The test is two-tailed.

2. $\alpha = \cdot10$

TEST SELECTION

1. The data are measured at the nominal scale.

2. N (the total population of each of the three *poviat*) is given in Table 5.1.

3. The requirements of the Poisson test appear to be met: brain tumours are rare and discrete events. As far as is known, this disease is non-contagious, hence the independence assumption should also be satisfied.

Table 5.1: Incidence of brain tumours in certain *poviat* in southern Poland

Poviat	Population (000s)	Numer of tumours per 100 000	Number of tumours Expected	Observed
Brzozow	70	0·0	3·619	0
Gorlice	83	10·843	4·291	9
Lesko	17	11·765	0·879	2

SOURCE: Choynowski (1959).

SAMPLING DISTRIBUTION

Brzozow. As no tumours were observed, we need to evaluate the probability of only one outcome, namely $p(0)$. $N = 70\,000$, $p = \cdot0000517$, hence $\lambda = NP = 3\cdot619$, and (using Table A, page 253)

$$p(0) = \frac{3\cdot619^0 \, e^{-3\cdot619}}{0!} = e^{-3\cdot619} = (e^{-3\cdot0})(e^{-0\cdot619})$$

$$= (\cdot04979)(\cdot5385) = \cdot0268$$

For illustrative purposes, the sampling distribution for $X = 0$ to $X = 9$ inclusive is shown in Figure 5.4A.

Gorlice. In this *poviat*, the observed number of tumours (9) exceeds the mean (4·291) hence the probabilities from 0 to 8 are required. Using the computational short cut,

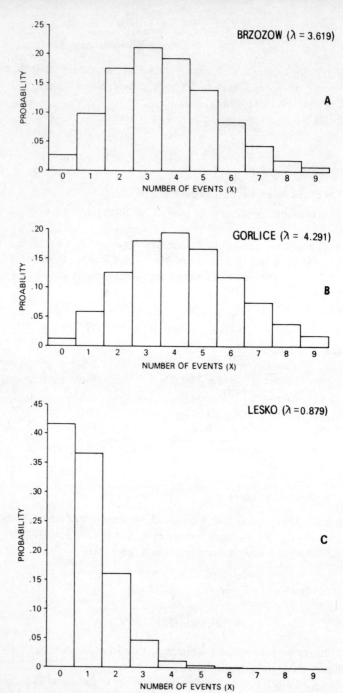

Figure 5.4: Sampling distribution for the range 0, 1, . . . , 9 for the incidence of brain tumours in 3 poviat in Poland (according to the Poisson distribution).

$$p(0) = \frac{4 \cdot 291^0 \, e^{-4 \cdot 291}}{0!} = e^{-4 \cdot 291} = \cdot 0137$$

$$p(1) = p(0)\frac{\lambda}{X} = \cdot 0137 \times 4 \cdot 291 = \cdot 0588$$

$$p(2) = \cdot 0588 \times \frac{4 \cdot 291}{2} = \cdot 1261$$

$$p(3) = \cdot 1261 \times \frac{4 \cdot 291}{3} = \cdot 1803$$

$$p(4) = \cdot 1803 \times \frac{4 \cdot 291}{4} = \cdot 1934$$

$$p(5) = \cdot 1934 \times \frac{4 \cdot 291}{5} = \cdot 1660$$

$$p(6) = \cdot 1660 \times \frac{4 \cdot 291}{6} = \cdot 1187$$

$$p(7) = \cdot 1187 \times \frac{4 \cdot 291}{7} = \cdot 0728$$

$$p(8) = \cdot 0728 \times \frac{4 \cdot 291}{8} = \cdot 0390$$

The sampling distribution for Gorlice is shown in Figure 5.4B.

Lesko. As in the case of Gorlice, more cases of brain cancer were diagnosed than expected, hence the critical region lies in the upper tail. As 2 cases were reported, the probabilities of 0 cases and of 1 case need to be evaluated.

$$p(0) = \frac{0 \cdot 879^0 \, e^{-0 \cdot 879}}{0!} = e^{-0 \cdot 879} = \cdot 4152$$

$$p(1) = p(0)\frac{\lambda}{X} = \cdot 4152 \times \cdot 879 = \cdot 3650$$

Figure 5.4C shows the sampling distribution for Lesko.

COMPUTATION

Brzozow. In this *poviat*, the expected number of cases of intracranial cancer is 3·619, but none were diagnosed; hence the critical region lies in the lower tail. The probability of none being diagnosed is ·0268 which is less than $\alpha/2$. Hence in this case we reject H_0 and conclude

that significantly fewer cases were diagnosed than was expected.
Gorlice. In this *poviat* nine cases of brain cancer were detected, and this
is more than double the expected frequency. We wish to know whether
the observed outcome lies in the critical region in the upper tail. The
probability of 9 or more persons contracting this disease is given by

$$p(9+) = 1 - \sum_{X=0}^{8} p(X) = 1 - \cdot9688 = \cdot0312$$

As this value is less than ·05 we again reject the null hypothesis.
Lesko. As in the case of Gorlice the observed frequency of brain cancer
was greater than expected and interest is focussed on the upper tail. Two
cases of brain cancer were reported therefore the probability of two or
more occurrences is

$$p(2+) = 1 - \sum_{X=0}^{1} p(X) = 1 - \cdot7802 = \cdot2198$$

The probability of two or more events far exceeds ·05, hence we must
accept H_0 in this case.

THE EXAMPLE IN SUMMARY

This example brings out very clearly the difference between frequencies
and probabilities; despite the fact that the number of tumours per
100 000 people is greater in Lesko than in Gorlice, the incidence of
tumours is significantly higher than in the population as a whole in
Gorlice, but not in Lesko. This apparent paradox is due to the fact
that in a *poviat* with a small number of residents, a high per caput
incidence can occur by chance more easily than in a poviat with a large
number of residents: there are almost five times as many residents in
Gorlice than in Lesko, and herein lies the answer to the paradox, and
the reason for mapping probabilities rather than absolute frequencies.

Geographical applications

As was the case with the binomial test, geographers face the problem
of identifying classes of events with sampling distributions which con-
form to the Poisson law. In the case of the Poisson distribution, several
appropriate classes of events have been identified. The example discussed
above using Choynowski's study of brain tumour incidence is one of a
number of possible applications in medical geography. For instance,

White (1971) has used the model to examine the distribution of leukemia mortalities in England and Wales. The model is not applicable to the study of contagious diseases (this would violate the independence assumptions) but there are a number of non-contagious diseases that can be examined using Poisson probability maps. White (1970) has also prepared probability maps of homicides and suicides. No doubt there are many other useful applications.

Another major application of the Poisson distribution in geography is in the study of point patterns. Suppose that one is examining a map with dots marked on it: these dots may represent anything that can be generalized as points on a map, such as places struck by lightning, or the location of farmsteads, or artesian wells. A whole host of phenomena occur at discrete points in space which can be treated as point patterns. On the mapped area, lay a regular grid of an appropriate size so as to create quadrats, and count the number of points in each quadrat. If the distribution of points is random, then the frequency of points in quadrats will correspond closely with the Poisson distribution.

The study that will be discussed in demonstrating the quadrat method is Clarke's (1948) work on the distribution of flying bombs that landed in South London during the Second World War. Within the study area, some quadrats were untouched by flying bombs, while in others there were several hits. Clarke's concern was whether the distribution of flying bombs provided statistical evidence that people living in certain areas were at greater risk than others; alternatively, the uneven distribution of flying bombs could have occurred by chance. The study area was divided into 576 quadrats, each measuring ¼ kilometre square. 537 flying bombs hit the study area, hence the mean expected number of bombs per quadrat is $537 \div 576 = \cdot932 = \lambda$. The probability of a cell being hit by X flying bombs is then evaluated in the usual way: for instance

$$p(0) = \frac{\cdot932^0 \, e^{-\cdot932}}{0!} = \cdot3938$$

The total expected number of quadrats with a frequency X is obtained by multiplying the probability of X by the total number of quadrats. Thus in Clarke's example, the expected number of quadrats with 0 flying bombs is $\cdot3938 \times 576 = 226\cdot7$. Clarke compared the frequency of cells in which 0, 1, 2, ... bombs fell to the expected frequencies under the Poisson distribution and found a very close correspondence, as Table 5.2 demonstrates.

Table 5.2: The observed and expected frequencies of quadrats hit by X flying bombs

Number of flying bombs landing in a quadrat (X)	Observed number of quadrats	Expected number of quadrats
0	229	226·7
1	211	211·4
2	93	98·5
3	35	30·6
4	7	7·1
5 or more	1	1·6
Total	576	575·9

SOURCE: Clarke (1948).

Although considerable use has been made of the quadrat method in plant ecology, the number of applications in geography is more limited. A study by Getis (1964) of changes in the distribution of grocery stores in Lansing, Michigan, will be used to raise two problems that arise with the quadrat method. First, the conclusions are only valid for the size of quadrat that is used. Getis analysed three different sizes of quadrat, and found fairly consistent trends at each scale, namely a trend for grocery stores to become more uniformly spaced. Although in nature one tends to find fairly consistent results, it is possible for different types of points patterns to be detected using different sizes of quadrat in the same area.

The second problem is deciding whether the Poisson model provides a good description of the process producing a point pattern. It would make little sense to compare a contagious point process such as pioneer settlement patterns, or persons afflicted with measles to the Poisson model because the independence assumption is clearly violated. Harvey (1966) has examined this methodological problem and pointed to one further problem, namely that two quite different point processes can give rise to very similar point patterns on the ground. Clearly it is much more meaningful to apply the Poisson model in situations where one comes close to meeting the independence assumptions: it is doubtful whether these assumptions are met in Getis's study. Nevertheless even in situations where the assumptions are violated the test is useful as an initial indicator of the character of the point pattern.

It would be premature to delve into the complexities of point pattern analysis at this juncture, for the main purpose is to introduce some of the geographical applications of the Poisson distribution. King

(1969, Chapters 3 and 5) discusses these complexities at greater length and also deals with a quite different method of point pattern analysis known as the nearest neighbour method which is not covered in this book.

One final problem arises. Suppose that one has an observed and expected frequency of quadrats with 0, 1, 2, . . . observations in them, as in Table 5.2. If they are very similar, then no further analysis is needed, but if they are dissimilar then one needs a rigorous method of comparing them. These methods, known as goodness-of-fit tests, are discussed in the next chapter.

6 Two tests of fit and association: the chi square and Kolmogorov–Smirnov tests

In some early textbooks, inferential statistics were referred to as 'sample statistics' since, for the most part, they are concerned with the relationships between samples, and between populations and samples. These relationships may be measured in a variety of ways, some simple, some quite complicated. The tests of fit and association discussed in this chapter are very simple tests: two are presented, the chi square test and the Kolmogorov–Smirnov test. The chi square test is probably the most frequently used of all statistical tests, while the Kolmogorov–Smirnov test is in common, but less frequent, use due to its more demanding assumptions. There are two versions of both of these tests. A one-sample test compares a sample to a specified theoretical population and tests how good is the correspondence or 'fit' between these two distributions – hence the term 'a test of fit'. The second version tests for association between or amongst two or more samples.

The one-sample chi square test

Statistical applications

The one-sample chi square test is used to compare a single sample with an 'expected' distribution. The sample consists of observations placed into two or more mutually exclusive categories so that the sample frequencies in each category can be compared to the frequencies given by the expected or theoretical distribution. The choice of an appropriate theoretical distribution will depend upon the type of expectations given in the research hypothesis. The simplest case (used in the example below) is to assume a uniform distribution; Clarke, in the example discussed at the end of the previous chapter, based his expected frequencies of flying bombs in quadrats on the Poisson distribution; any one of a number of other distributions may also be used.

The one-sample chi square test is concerned with the *magnitude* of the differences between a sample and the theoretical distribution. Figure 6.1 shows a typical sampling distribution for chi square. The test shown

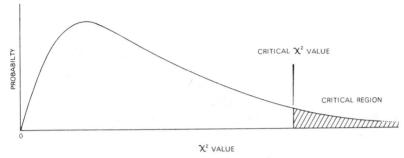

Figure 6.1: A typical sampling distribution for chi square showing the location of the critical region in a test for significant differences.

in the figure is the conventional test for significant differences: normally one hopes that the chi square value, which is based on the magnitude of the summed differences between the sample and the population, is so large that it is not likely to have occurred by chance. As a result, the critical region conventionally lies in the upper tail of the sampling distribution. Occasionally the chi square test is used to test for significant similarities, in which case the critical region lies in the lower tail.

Assumptions

The data may be measured at the nominal scale or any higher level of measurement.

There must be at least two mutually exclusive categories into which the observations are placed.

When there are only two categories, the expected frequency in each category must not be less than 5.

When there are more than two categories, no category should have an expected frequency less than 1, and not more than one category in five should have an expected frequency of less than 5. Where this assumption is not met, it may be possible to amalgamate adjacent categories to bring the expected frequencies up to the required level.

Test statistic

Let k represent the number of categories,
        O_i be the observed number of cases in category i,
and E_i be the expected number of cases in category i:
then the chi square value (χ^2) is given by

$$\chi^2 = \sum_{i=1}^{k} \frac{(O_i - E_i)^2}{E_i} \qquad \qquad 6.1$$

It is important to distinguish between the computed chi square value that is given by the above formula, and the expected chi square value that is given by the chi square distribution. Conceptually these two values are quite distinct: the former is obtained empirically by manipulating data; the latter is obtained from a theoretical sampling distribution which closely approximates the distribution of χ^2 under the null hypothesis. To be exact, there is a family of theoretical chi square distributions, each one having its precise shape determined by the number of degrees of freedom (represented by the Greek letter ν – pronounced 'nu'). In the one sample test the number of degrees of freedom is given by

$$\nu = k - 1 \qquad \qquad 6.2$$

There are $(k - 1)$ degrees of freedom for the following reason: suppose that 30 observations are to be placed in three categories: if 12 observations are placed in category one, and 8 in category two, then the number in the third category is determined; in order for the total to equal 30, there *have* to be 10 observations in the third category. In other words, there are two degrees of freedom with three categories: there is freedom in placing the first two values, but no further freedom if the total in the three categories is to equal N.

Critical values of chi square, given under the null hypothesis, are to be found in Table B, page 254. The left-hand column indicates the number of degrees of freedom, the other columns the level of significance. Hence if one has 4 degrees of freedom and $\alpha = \cdot 01$, then one goes across the row for $\nu = 4$ until one reaches the column headed $\alpha = \cdot 01$, and the critical chi square value is 13·28. The chi square test involves comparing the calculated χ^2 value to χ_α^2 given in the table: for the usual test of significant differences, if $\chi^2 \geqslant \chi_\alpha^2$ one rejects H_0, but if $\chi^2 < \chi_\alpha^2$ one must accept H_0. (For the test of significant similarities these relationships are reversed.)

Before presenting an example, one further digression is needed. Many textbooks advocate the use of a *correction for continuity,* known as Yates correction, with the chi square test. The reason for doing this is that the theoretical distribution assumes continuous values, whereas the observations that form the basis of χ^2 occur as integers. There are two reasons for not advocating the use of Yates correction here. First, as Croxton, Cowden and Klein (1968, p. 594) point out, the correction overcorrects so that the 'corrected' χ^2 value can be almost as far from the true value as the uncorrected value is. Second, the correlation coefficient based on chi square (known as the phi coefficient) cannot attain its theoretical limits of +1 and −1 if Yates correction is applied.

Example

Archaeological investigations over many years have built up a considerable amount of evidence on the location of Huron Indian villages in Oro township in Central Ontario. This example explores the relationship between Huron village location and topography: in particular we wish to discover whether there is a significant association between village location and local slope conditions.

Figure 6.2 shows the location of the known village sites and the slope characteristics of Oro township. The percentages in Table 6.1 indicate a preference for rolling ground since more than half of the villages are located on the 36·4% of the land area classified rolling and steep. The one sample chi square test may be used to examine the possibility that this relationship did not occur by chance.

Table 6.1: Location of Huron villages in Oro township, Ontario

	Percentage of total area	Number of villages Expected	Observed
Flat land	63·6	31·8	22
Rolling and steep land	36·4	18·2	28
Total	100	50	50

PROBLEM DEFINITION

1. H_0: There is no significant association between the location of Huron villages and slope conditions.

H_1: There is a significant relationship between the location of Huron Indian villages and local slope conditions (which are dichotomized as flat, or rolling).

2. $\alpha = \cdot05$

TEST SELECTION

1. The data are measured at the ordinal scale: some villages are located

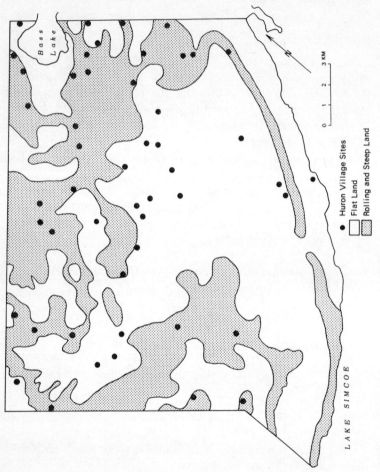

Figure 6.2: Slope characteristics and Huron Indian village location in Oro township, Ontario.
SOURCE: C. E. Heidenreich, personal communication.

in areas with little local relief ('flat' land) others in areas with more local relief ('sloping' terrain).

2. $N = 50$; there are 22 observations in the first category and 28 in the second.

3. The one-sample chi square test is an appropriate test: the frequency assumptions are met.

SAMPLING DISTRIBUTION

As there are two categories, the number of degrees of freedom is

$$\nu = k - 1 = 2 - 1 = 1$$

Consulting Table B (page 254) the critical chi square value for $\nu = 1$ and $\alpha = \cdot05$ is 3·84. The computed chi square value must be at least as large as the critical value if we are to reject the null hypothesis.

COMPUTATION

The calculations are based on the data in Table 6.1. The expected values in this table are based on the assumption under the null hypothesis, that Huron villages are uniformly distributed per unit area: ·636 of the total area of the township is classified 'flat', ·364 is classified 'rolling' and these proportions are multiplied by the total number of villages (50) to obtain the expected values.

Using the formula

$$\chi^2 = \sum_{i=1}^{k} \frac{(O_i - E_i)^2}{E_i} \qquad \textbf{6.1}$$

we have

$$\chi^2 = \frac{(22 - 31 \cdot 8)^2}{31 \cdot 8} + \frac{(28 - 18 \cdot 2)^2}{18 \cdot 2} = \frac{96 \cdot 04}{31 \cdot 8} + \frac{96 \cdot 04}{18 \cdot 2} = 3 \cdot 04 + 5 \cdot 28 = 8 \cdot 32$$

This value is considerably larger than the critical value, and in consequence H_0 is rejected. We conclude that there is a significant association between village location and local topography.

Geographical applications

These are considered after discussion of the chi square test for two or more samples.

The chi square test for two or more independent samples

Statistical applications

Conceptually, this test is quite similar to the one sample chi square test, although in this case the comparison is made between two, or more, independent samples. The samples consist of *absolute* frequency data (i.e. percentages or other relative frequencies are not suitable), and each sample is split into the same k mutually exclusive categories. By comparing the sample frequencies in each category, one can determine the probability that the samples were drawn from different populations.

Assumptions

The data may be measured at the nominal scale or any higher level of measurement.

There must be at least two samples and at least two mutually exclusive categories into which the observations are placed.

No category should have an expected frequency less than 1, and not more than one category in five should have an expected frequency less than 5.

Test statistic

Let k = the total number of categories
       ℓ = the total number of samples
       O_{ij} = the observed frequency in category i of sample j
       E_{ij} = the expected frequency in category i of sample j
Then the test statistic is given by

$$\chi^2 = \sum_{i=1}^{k} \sum_{j=1}^{\ell} \frac{(O_{ij} - E_{ij})^2}{E_{ij}} \qquad \textbf{6.3}$$

In the one-sample chi square test, the expected frequencies are given by the theoretical distribution, be it normal, Poisson or whatever. In the present case with two or more independent samples, for any given category in any given sample, the expected frequency is obtained as follows: one multiplies the total number of observations in *all* samples in that given category by the total number of observations in *all* categories in the given sample, and divides the result by N (the total of all observations in all samples). This method of obtaining estimated frequencies is illustrated in the example that follows.

When the chi square value has been calculated according to formula **6.3**, it is tested for significance by comparing the computed value to the critical value given in Table B (page 254) for the required level of significance (α) and degrees of freedom (ν). The number of degrees of freedom is given by

$$\nu = (k-1)(\ell-1) \qquad\qquad \textbf{6.4}$$

where, as before, k is the number of categories and ℓ the number of samples. For the usual test where one is looking for significant differences, a χ^2 value is deemed significant when it is larger than the tabled value (i.e. when $\chi^2 \geqslant \chi_\alpha^2$).

Example

Data for the English city of Plymouth will be used to examine whether the industrial composition of this town varies systematically from one part of the city to another. For the purposes of this test, the city will be divided into three areas, the docks and adjacent areas ('dockside'), the older parts of the city that are not adjacent to the docks ('centre'), and the newer suburban areas ('suburbs'). The number of manufacturing establishments employing more than ten people in 1964 in each of these three areas was then tabulated, using the following four-fold classification:

1. Primary industries plus shipbuilding: these are the types of activities that one would expect to be attracted to a dock area.

2. Engineering, brick and cement industries: these industries might be concentrated in suburban areas, either because they are newer growth industries or because they are big space users.

3. Textiles, leather, clothing and footwear.

4. Baking, furniture and other wood using industries, paper, printing and publishing, and other miscellaneous manufacturing industries: some of these are long established activities that might be concentrated in the central parts of the city.

The observed and expected frequencies are given in Table 6.2. The expected frequencies were obtained by multiplying the respective row and column totals and dividing by the grand total. Thus for activity group 2 for the suburbs the expected frequency (E_{23}) is

$$E_{23} = \frac{35 \times 31}{125} = 8.7$$

Table 6.2: Manufacturing establishments in Plymouth in 1964, classified by location and activity group (expected frequencies are given in brackets)

Activity group	Dockside	Centre	Suburbs	Total
1	24	8	7	39
	(18·1)	(11·2)	(9·7)	
2	11	9	15	35
	(16·2)	(10·1)	(8·7)	
3	7	4	3	14
	(6·5)	(4·0)	(3·5)	
4	16	15	6	37
	(17·2)	(10·7)	(9·2)	
Total	58	36	31	125

SOURCE: data collected by author.

PROBLEM DEFINITION

1. H_0: the industrial composition does not vary significantly amongst the three areas in Plymouth.

H_1: there are significant variations in industrial composition amongst the three areas.

2. $\alpha = \cdot 05$.

TEST SELECTION

1. Measurement is at the nominal scale.

2. $N = 125$. All the expected frequencies are greater than 1, while only two are less than 5 and this is fewer than one category in five.

3. the assumptions appear to be met for the chi square test for k independent samples.

SAMPLING DISTRIBUTION

Under the null hypothesis, χ^2 is distributed approximately as chi square with $(k-1)(\ell-1)$ degrees of freedom. In the present case

$$\nu = (4-1)(3-1) = 6$$

and consulting Table B (page 254), $\chi^2_{\cdot 05} = 12\cdot 59$.

COMPUTATION

The simplest way to arrange the computations with this type of problem is to draw up a table as in Table 6.3. One can then work out the results for each row in the table, and sum the final column to obtain the chi square value: in the present example $\chi^2 = 12\cdot98$. By using this tabular form, one can apply two simple checks to one's calculations: first, the sum of the expected values should match the sum of the observed values; and second the sum of the differences, $O_{ij} - E_{ij}$, should equal zero (subject to rounding errors).

Table 6.3: Calculating a chi square value

Category	Sample					$\dfrac{(O_{ij} - E_{ij})^2}{E_{ij}}$
(i)	(j)	O_{ij}	E_{ij}	$(O_{ij} - E_{ij})$	$(O_{ij} - E_{ij})^2$	
1	1	24	18·1	5·9	34·81	1·92
1	2	8	11·2	−3·2	10·24	0·91
1	3	7	9·7	−2·7	7·29	0·75
2	1	11	16·2	−5·2	27·04	1·67
2	2	9	10·1	−1·1	1·21	0·12
2	3	15	8·7	6·3	39·69	4·56
3	1	7	6·5	0·5	0·25	0·04
3	2	4	4·0	0·0	0·00	0·00
3	3	3	3·5	−0·5	0·25	0·07
4	1	16	17·2	−1·2	1·44	0·08
4	2	15	10·7	4·3	18·49	1·75
4	3	6	9·2	−3·2	10·24	1·11
Totals		125	125·1	−0·1		12·98

The results indicate that the chi square value of 12·98 slightly exceeds the critical value of 12·59, hence we reject the null hypothesis and conclude that the mix of industry does vary significantly amongst the three areas in Plymouth.

Geographical applications

The range of applications of the chi square test is so large that it would not be very informative to present a list — indeed such a list is unnecessary as many applications are self-evident. Comment will therefore be confined to one general application that has not been referred to by example, and one important geographical limitation.

The chi square test may be used as a screening mechanism when one has many observations on a large number of variables which one plans to use in multivariate analysis. Typically this type of data is obtained from interview surveys, or from census authorities (particularly now that census information is becoming available on magnetic tapes). Given this type of data, chi square tests are applied through pairwise comparisons of variables, so that an overview of interrelationships between the variables is obtained. This type of cross-tabulation or multivariate counting procedure fulfils a diagnostic role, while making only minimal assumptions about the form of the data that is used.

An important limitation of the chi square test is the need for absolute frequency data: relative frequency data, such as percentages, distances, and areas should not be used. It was this confusion that led to a debate between Zobler and Mackay in the late 1950s. Zobler reported on a study in which he conducted chi square tests on data recording the number of acres devoted to different forms of land use. Mackay pointed that by changing one's basis of measurement to a different size of areal unit, one obtained an entirely different chi square value although the number of degrees of freedom remains unchanged.

The Kolmogorov–Smirnov one-sample test

Statistical applications

Suppose that one wishes to compare a sample consisting of data classified into k categories with a theoretical distribution, and the categories are ordered from smallest to largest, or vice versa. The χ^2 test may be used, but another test known as the Kolmogorov–Smirnov test is frequently more appropriate. Properly, there are two closely related tests: Smirnov's test is concerned with the *summed* difference between a theoretical and an empirical cumulative distribution; Kolmogorov's test (which was extended by Smirnov and subsequently became known as the Kolmogorov–Smirnov test) is concerned with the *maximum* difference between two such distributions. The distribution theory for Smirnov's statistic is difficult and the test is not presented here.

Like the chi square test, Kolmogorov's statistic may be used both to compare an empirical to a theoretical frequency distribution, or to compare two empirical distributions with one another. However, the Kolmogorov–Smirnov (K–S) test has a more limited range of applications in two important ways: it requires data measured at the ordinal scale; and it cannot be used to compare more than two samples in a single test. To counter these limitations, the Kolmogorov–Smirnov test has two advantages over the chi square: first, there are no limitations on the size of sample and frequencies required in categories; and second, the K – S test is usually more powerful (although Siegel (1956, p. 56 and p. 136) is wrong in asserting that it is always more powerful).

Assumptions

Observations should be independent events measured at the ordinal scale or at a higher level of measurement.

There must be at least two mutually exclusive categories into which the observations are classified.

The population from which a sample is drawn should have a continuous distribution. It would seem, however, that this assumption can be safely relaxed: Goodman (1954) has found the K – S test to be conservative when discrete data is used i.e., the probability of making a Type II error – of accepting H_0 when it should be rejected – is slightly increased.

Test statistic

A cumulative frequency distribution expresses the proportion of the observations in a sample that are less than or equal to a given value. It is obtained by ranking the observations from smallest to largest, and then summing them together and expressing the result as a proportion of N, so that the cumulative distributions range from 0 to 1·0. Cumulative distributions may be expressed as fractions or decimals, whichever is the more convenient.

Two cumulative distributions are required for this test: let

$F_0(X)$ = a theoretical cumulative distribution function that is given under H_0. The 'expected' values are given by the theoretical distribution.

$S_N(X)$ = an empirical cumulative distribution function.

The Kolmogorov–Smirnov test relates to the *maximum difference, D,* between these two cumulative distributions. Hence

$$D = \max |S_N(X) - F_0(X)| \qquad \textbf{6.5}$$

The vertical lines in equation **6.5** indicate that the absolute value of the difference is required, and negative signs are ignored.

The sampling distribution for D under H_0 for two-tailed tests is given in Table C, page 255 (tables are not available for one-tailed tests). Clearly, one would expect some deviation between an observed and a theoretical distribution as a result of sampling variations, hence rejection of the null hypothesis depends upon the deviations being large. H_0 is rejected when D is equal to or greater than the critical value, D_α, given in Table C.

Example

Under most circumstances, the K – S test is a useful way of examining the distribution of points in quadrats. In the present case, the test is

Figure 6.3: The distribution of towns in Gippsland, Australia.
SOURCE: G. Robinson and K. J. Fairbairn (1969).

applied to the distribution of 154 towns in Gippsland, Australia. A grid was placed over the map in Figure 6.3 and, with some amalgamation of adjacent and partly occupied quadrats at the edge of the map, the area was covered with 158 quadrats. The frequency of quadrats with 0, 1, 2, 3, 4, 5, 6 and 7+ towns was then counted (see Table 6.4).

Since the present test is to determine whether the towns in Gippsland are randomly distributed, the Poisson distribution will serve as the expected distribution under the null hypothesis.

PROBLEM DEFINITION

1. H_0: the distribution of towns in Gippsland does not differ significantly from a random distribution.

H_1: the distribution of towns in Gippsland is non-random.

2. $\alpha = \cdot05$

Table 6.4:   The observed and expected frequencies of towns in quadrats in Gippsland, Australia

Number of towns in a quadrat	Probability under the Poisson model	Expected number of quadrats	Observed number of quadrats
	(A)	(A × 158)	
0	·3773	59·61	85
1	·3677	58·10	36
2	·1792	28·31	16
3	·0582	9·20	9
4	·0142	2·24	6
5	·0028	0·44	2
6	·0004	0·06	3
7+	·0002	0·03	1
Total	1·0000	157·99	158

TEST SELECTION

1. The data is measured at the ordinal level.

2. $N = 158$.

3. A one sample K–S test comparing the observed distribution to the distribution expected under the Poisson model is appropriate.

SAMPLING DISTRIBUTION

The critical values of the test statistic, D, are given in Table C, page 255. For $N = 158$ we have

$$D_{\cdot05} = \frac{1\cdot36}{\sqrt{158}} = \cdot108$$

If the maximum deviation between the observed and theoretical cumulative distributions is greater than ·108, then H_0 will be rejected.

COMPUTATION

Once the expected frequencies have been evaluated, the next step is to calculate the cumulative distributions $F_0(X)$ and $S_{158}(X)$. It is easier to do the calculations in tabular form, as in Table 6.5, and in this case the cumulative distributions are expressed as fractions.

Table 6.5:   Kolmogorov–Smirnov one sample test applied to the distribution of towns in quadrats for Gippsland, Australia

	Number of towns in quadrats							
	0	1	2	3	4	5	6	7+
$S_{158}(X)$	$\dfrac{85}{158}$	$\dfrac{121}{158}$	$\dfrac{137}{158}$	$\dfrac{146}{158}$	$\dfrac{152}{158}$	$\dfrac{154}{158}$	$\dfrac{157}{158}$	$\dfrac{158}{158}$
$F_0(X)$	$\dfrac{59 \cdot 61}{158}$	$\dfrac{117 \cdot 71}{158}$	$\dfrac{146 \cdot 02}{158}$	$\dfrac{155 \cdot 22}{158}$	$\dfrac{157 \cdot 46}{158}$	$\dfrac{157 \cdot 90}{158}$	$\dfrac{157 \cdot 96}{158}$	$\dfrac{157 \cdot 99}{158}$
$\lvert S_{158}(X) - F_0(X) \rvert$	$\dfrac{25 \cdot 39}{158}$	$\dfrac{3 \cdot 29}{158}$	$\dfrac{9 \cdot 02}{158}$	$\dfrac{9 \cdot 22}{158}$	$\dfrac{5 \cdot 46}{158}$	$\dfrac{3 \cdot 90}{158}$	$\dfrac{0 \cdot 96}{158}$	0

Table 6.5 indicates that $D = \dfrac{25 \cdot 39}{158} = \cdot 161$ which is larger than

$D_{\cdot 05}(\cdot 108)$; hence we reject H_0 and conclude that towns in Gippsland are not randomly distributed.

In this case, the investigation can be taken a little further to examine the nature of the departure from randomness. There are more quadrats than expected with 0, 4, 5, 6 and 7 towns in them, and fewer with 1, 2 and 3 towns. Insofar as the expected distribution is a random distribution, then the deviations from randomness are towards a clustered point pattern: more cells than expected are either vacant or are occupied by an above average number of points.

It was suggested earlier that, contrary to some statements, the K–S test is not always more powerful than chi square. This example provides some indication why this is so: if one examines the individual differences between the observed and the theoretical distribution for the eight categories from 0 to 7+, they are (by sign) +, −, −, −, +, +, +, +. The first discrepancy is the largest, and the next three differences are of the opposite sign so that the cumulative difference diminishes. Quite frequently with point patterns the individual differences tend to alternate +, −, +, −, + . . . so that successive differences tend to cancel each other out and the cumulative difference never becomes large. In several empirical examples where this situation occurred, the chi square test has revealed significant differences where the K–S test has failed to establish significance.

Geographical applications

These are discussed after the two sample K–S test is presented.

The Kolmogorov–Smirnov test for two independent samples

Statistical applications

This test is similar to the equivalent one sample test, except that attention is focused on the maximum difference between two empirical cumulative distributions. The test is used to determine whether two samples are drawn from the same or different populations. Unlike the one sample test, both one and two-tailed tests may be performed in the case of the two sample test.

Assumptions

These are the same as for the one sample test.

Test statistic

Let m and n be the number of observations in Sample I and Sample II respectively, and $N = m + n$. Using $S_m(X)$ and $S_n(Y)$ to represent the two empirical cumulative distribution functions, then the two-tailed test is concerned with the maximum absolute difference between $S_m(X)$ and $S_n(Y)$; i.e.,

$$D = max \mid S_m(X) - S_n(Y) \mid \qquad \textbf{6.6}$$

The one-tailed test, in contrast, is based on the maximum difference in a specified direction between two cumulative distributions, or

$$D = max \, (S_m(X) - S_n(Y)) \qquad \textbf{6.7}$$

The test criterion for the two-tailed test involves comparing D to the critical value of D_α given in Table D, page 256. When $D < D_\alpha$, one concludes that the two samples are drawn from populations with the same distribution and the small differences are due to random effects: however when $D \geqslant D_\alpha$ then the difference is deemed to be 'real', and one concludes that the two samples are drawn from populations with different distributions.

The test criterion for the one-tailed test makes use of the χ^2 distribution (Table B, page 254). Having calculated D, one evaluates

$$\chi^2 = 4D^2 \frac{mn}{m+n} \qquad\qquad 6.8$$

and compares this to χ_α^2 with 2 degrees of freedom: H_0 is rejected when $\chi^2 \geqslant \chi_\alpha^2$.

Example

The basic tenet of Weberian location theory may be briefly stated as: when the manufacturing process involves weight loss in converting raw materials into a product, then industry is attracted to raw material sources. On the other hand when no weight loss is involved, or when ubiquitous materials such as water are added to localized raw materials so that a gain in weight takes place, then industry is not attracted to the raw material location (unless this happens to coincide with the market).

Wilfred Smith (1955) has examined the validity of this Weberian proposition using data from the British Census of Manufacturing for 1948. In this case there are two samples:

1. Industries located wholly or partly at raw material sources
2. Industries not located at raw material sources.

Smith classified the various industries according to their Material Index (a measure of weight loss during manufacture) using 11 categories ranked from A — with considerable weight loss — through to K — where an increase in weight takes place. The resulting data are given in Table 6.6.

Table 6.6:   Industry classifed by type of location and amount of weight loss

Industry	Material index											Total
	A	B	C	D	E	F	G	H	I	J	K	
Located wholly or partly at raw materials (X)	2	4	1	3	4	4	4	6	3	3	0	34
Not located at raw materials (Y)	0	3	3	2	1	6	2	2	1	3	8	31

SOURCE: Based on Smith (1955) Table 1.

PROBLEM DEFINITION

1. H_0: there is no significant difference between the material indices for industries located and not located at raw material locations.

H_1: industries located wholly or partly at raw material locations tend to involve greater weight loss and have higher material indices than do industries not located at raw material locations.

The test is one-tailed.

2. $\alpha = \cdot05$

TEST SELECTION

1. The data are placed in categories that are ranked at the ordinal scale with A representing the highest material index and K the lowest.

2. m(sample I) = 34

n (sample II) = 31

$N = m + n = 65$

3. The conditions are met for a two sample Kolmogorov–Smirnov test.

SAMPLING DISTRIBUTION

The sampling distribution of the test statistic, D, is closely approximated by chi square with two degrees of freedom. For $\nu = 2$ and $\alpha = \cdot05$, the critical value in Table B, page 254, is 5·99.

COMPUTATION

As in the one sample K–S test, the easiest way to present the calculations is by way of a table. In Table 6.7, the two cumulative frequency distributions $S_{34}(X)$ and $S_{31}(Y)$ are given in decimals, together with the differences between the two cumulates. The largest difference in the postulated direction is found in category I, hence $D = \cdot273$. Evaluating equation **6.8**, we have

$$\chi^2 = 4(\cdot273)^2\left(\frac{1054}{65}\right) = 4\cdot834$$

Table 6.7: Kolmogorov–Smirnov two sample test applied to the location of industry in Britain, 1948

	A	B	C	D	E	F	G	H	I	J	K
$S_{34}(X)$	·059	·176	·206	·294	·412	·529	·647	·824	·918	1·00	1·00
$S_{31}(Y)$	·000	·097	·194	·258	·290	·484	·548	·613	·645	·742	1·00
$(S_{34}(X) - S_{31}(Y))$	·059	·079	·012	·036	·122	·045	·099	·211	·273	·258	0·0

As $\chi^2 < \chi_\alpha^2$, we must in this case accept the null hypothesis. It is worth noting that the differences would be significant at the ·10 level, hence the results are sufficiently interesting to warrant further exploration of the problem.

Geographical applications

The most common application of the one sample K–S test in geography is in comparing either point patterns or distance frequencies to various theoretical distributions. The Gippsland example which was developed earlier in this chapter was used by Norcliffe (1969) to ascertain whether or not towns in this region are randomly distributed. Dacey (1968) has used a similar procedure to examine the distribution of houses in Puerto Rico: he compared the distribution of houses in quadrats to the negative binomial distribution (which describes a contagious process giving rise to a somewhat clustered distribution) and generally found a good correspondence. Morrill and Pitts (1967) present a number of examples in which migration and contact data (organized in distance bands) are compared to Pareto, exponential and Pareto-exponential distributions using the K–S test.

Somewhat similar to the Morrill and Pitts application is the work of Blaikie (1971) who used the K–S test in examining crop distribution patterns in fields around the Indian villages of Daiikera and Keru in Rajasthan. This study raises a problem that has already been mentioned in discussing geographical applications of chi-square — the problem of modifiable units. Clearly, fields are areally extensive, but are they modifiable units like acres and square yards? Blaikie was aware of this problem: he points out that since Kolmogorov's statistic is based on proportional frequencies, it is independent of the size of the unit of measurement. However, there is an ambiguity in this statement that needs clarifying: the *magnitude* of D is largely independent of the measurement unit, but the *significance* of D most certainly depends on the size of the sample, and therefore the unit of measurement. Suppose, for instance, that every field was divided in half: the value of D would remain the same, but m and n would be doubled and the value of χ^2 in the two sample test would be substantially altered. The debate on the definition of modifiable units has yet to be resolved, but in the meantime it is appropriate to be cautious in applying significance tests to data that are measured over modifiable units.

7 Nonparametric measures of correlation

One frequently finds statements to the effect that two things are 'correlated', the term being used loosely to describe an assumed relationship. Statisticians use the term *correlation* in a much more specific way to denote a measurable statistical relationship between two or more variables. There are several coefficients of correlation; the five discussed in this book are selected for their computational simplicity and utility. Three coefficients are presented in this chapter, the phi coefficient (ϕ), Spearman's coefficient of rank correlation (r_s), and the point biserial coefficient of correlation (r_{pb}); two are discussed later — the coefficient of medial correlation (q) in Chapter 8 and Pearson's product moment coefficient of correlation (r) in Chapter 11.

The phi coefficient

Statistical applications

The phi coefficient is a special case of a more general method which is known as the *correlation of attributes* since the categories into which observations are placed may be considered as attributes of the observations. The correlation of attributes may be applied to contingency tables of any size; the phi coefficient applies strictly to 2 x 2 contingency tables. This measure is one of a variety of coefficients that can be computed for contingency tables; Sokal and Sneath (1963, Chapter 6) discuss several others. The phi coefficient is included here, firstly because it is a logical extension of the χ^2 test, and secondly because it is very easy to compute.

A phi coefficient can be obtained directly from a χ^2 value, and vice versa, but whereas a χ^2 value may vary between 0 and N, ϕ ranges between -1 and $+1$. Given the ease with which a chi square test may be performed, it may seem puzzling that one should go to the effort of computing a phi coefficient. However, the two measures are quite different in character: the chi square test reveals information about the *significance* of an association between two variables, whereas the phi coefficient reveals information about the *strength* of that association. It is possible for an association to be significant, yet quite weak, just as a strong association may be insignificant it if is based on a small sample.

Assumptions

The assumptions are the same as for a chi square test with two samples and two categories.

Test statistic

The phi coefficient is computed for 2 x 2 contingency tables which may be set up as follows:

	Category I	Category II	Total
Sample 1	A	B	$A + B$
Sample 2	C	D	$C + D$
Total	$A + C$	$B + D$	N

Using the notation in this table, then phi (ϕ) is given by

$$\phi = \frac{AD - BC}{\sqrt{(A + B)(C + D)(A + C)(B + D)}} \qquad 7.1$$

The sign attached to ϕ is determined by the cross products in the numerator: the result is positive if $AD > BC$, and negative if $AD < BC$.

In some instances, a chi square value is already known, in which case ϕ can be obtained easily from

$$\phi = \sqrt{\frac{\chi^2}{N}} \qquad 7.2$$

There is one problem in obtaining ϕ from χ^2: the latter is always a positive value, hence it will always give rise to a positive ϕ value. In consequence, it is sometimes necessary, when using equation 7.2, to interpolate a negative value by visual inspection of the contingency table. Equation 7.1 is therefore recommended, and if χ^2 is needed, then

$$\chi^2 = \phi^2 N \qquad 7.3$$

Example

In 1972, an interview survey was conducted in Southern Bruce county on the eastern shores of Lake Huron, in Ontario. For many years this has been a stable farming area experiencing some net outmigration, especially of younger people. However in the late 1960s a radical change began as a result of development of a major nuclear generating complex in the area. This has reversed the pattern of outmigration; a large number of electrical engineers, construction workers and their families have been attracted to the area by the new employment opportunities.

This study focused on comparing the contact patterns of residents who had grown up in the area with those of the newly arrived residents who had grown up outside the local area. Interviewees were asked about several aspects of their contact patterns, for instance the frequency and distance of contacts with relatives, friends, retail outlets, places of work, clubs, churches and so on. The responses were cross-classified by age, place of birth, place of residence and income, thereby generating a large number of contingency tables. A number of interesting and significant relationships were identified.

Table 7.1   Place of birth and grocery shopping location for a sample of residents of Southern Bruce county, Ontario

	Trip Distance for major weekly grocery shopping		
	$\leqslant$25 km	> 25 km	Total
Born inside the study area	38	18	56
Born outside the study area	5	10	15
Total	43	28	$N = 71$

SOURCE: Data collected by the author.

This example takes up the relationship between an interviewee's place of birth (inside or outside the study area) and the distance travelled to do the household's major weekly grocery shopping. As Table 7.1 indicates, grocery shopping trips were classified into two categories, those of more than 25 kilometres, and those of shorter length. The reason for choosing these two categories is that in 1972, the study area lacked a major supermarket store, and trips of greater than 25 kilometres had to be made in order to reach such a store. Were people living in the study area, but born outside it, more likely to undertake fairly long trips in order to shop at a large supermarket? This question is explored using data which, as far as possible, holds other variables such as age and income constant. The table is based on responses from housewives aged between 20 and 40 years; income distribution is quite similar for those born inside and outside the study area for this age group.

PROBLEM DEFINITION

1. H_0: there is no association between the distance that people travel to do their weekly grocery shopping and the birthplace of an interviewee (inside or outside the study area).

H_1: the distance travelled to do weekly grocery shopping is associated with the birthplace of an interviewee.

2. $\alpha = \cdot 05$

TEST SELECTION

1. Both variables are measured at the ordinal level. The distance variable involves two categories: trip less than or equal to 25 kilometres, and trips greater than 25 kilometres. The place of birth variable divides interviewees into those born inside the study area, and those born outside (and therefore at a greater distance from) the study area.

2. $N = 71$.

3. The data is classified by a 2 x 2 contingency table suitable for the computation of ϕ.

SAMPLING DISTRIBUTION

In this case, ϕ describes the strength of association between place of birth and length of grocery shopping trip. However the coefficient may be tested for significance by converting ϕ into χ^2 and comparing to χ^2_α with 1 degree of freedom ($\chi^2_{\cdot 05} = 3.84$)

COMPUTATION

From **7.1** we have

$$\phi = \frac{(38 \times 10) - (18 \times 5)}{\sqrt{(56 \times 15 \times 43 \times 28)}} = \frac{290}{1005 \cdot 7} = \cdot 288$$

A ϕ of $\cdot288$ indicates a fairly weak positive association. Checking this value for significance using **7.3**, we have

$$\chi^2 = (\cdot288)^2 \times 71 = 5\cdot90$$

which exceeds the critical value of $3\cdot84$. This indicates that there is a weak but significant positive association between a person's place of birth and length of weekly grocery shopping trip.

Geographical applications

The phi coefficient is applicable in any situation where a 2 x 2 contingency test is legitimate. An interesting geographical example is to be found in a paper by W. S. Robinson (1950) who correlated race with literacy in the United States. Using the individual data given in Table 7.2, Robinson found $\phi = \cdot203$, indicating a weak positive association between colour and literacy. This paper raises the important problem of ecological correlations which are discussed in Chapter 13.

Table 7.2: Phi coefficient correlating colour and literacy for the United States, 1930*

	Black	White	Total
Illiterate	1 512	2 406	3 918
Literate	7 780	85 574	93 354
Total	9 292	87 980	97 272

$$\phi = \cdot203$$

*The figures are in thousands, and exclude people under ten years of age.

SOURCE: Robinson (1950, p. 353).

Spearman's coefficient of rank correlation

Statistical applications

Ranked data has the characteristic that the interval between successive observations is fixed at unity, except in the case of ties. Thus the following seven observations (measured at the interval level)

4.7   9.6   2.1   3.3   9.6   10.4   6.6

after being converted to the rank scale and ranked from smallest to largest become

$$3, \quad 5\cdot5=, \; 1, \quad 2, \quad 5\cdot5 =, \; 7, \quad 4$$

For data of this type, two correlation coefficients are available — Kendall's rank correlation coefficient (known as Kendall's τ (tau)) and Spearman's rank correlation coefficient (r_s). Both measures are equally powerful in rejecting H_0, although Spearman's coefficient has the edge in terms of conceptual and computational simplicity, and for this reason is presented here. Details of Kendall's τ may be found in Siegel (1956, Chapter 9).

Spearman's r_s is a measure of linear correlation between two variables, X and Y, which means that if the correlation was plotted on a graph of X and Y, it would appear as a straight line. r_s displays the two characteristics desired of a correlation coefficient:

1. It ranges between the limits of +1 (indicating a perfect positive relationship) and −1 (indicating a perfect negative relationship).
2. It has a value of 0 when the two variables are totally unrelated.

The sample correlation coefficient, r_s, is used to make inferences about the population correlation coefficient, ρ_s (rho). The research hypotheses that are most commonly applied are: $\rho_s \neq 0$ (a two-tailed test), and $\rho_s > 0$ or $\rho_s < 0$ (a one-tailed test).

Assumptions

1. Observations should be ranked in order of magnitude on both variables, X and Y.
2. There must be at least five pairs of observations to establish significance at a generally meaningful level.

Test statistic

Let N be the number of observations

          X be the rank of an observation on the first variable

          Y be the rank of an observation on the second variable

and d be the difference between these two rankings.

Thus if an object is ranked 1 on X, and 4 on Y, then $d = 1 - 4 = -3$.

Then Spearman's coefficient of rank correlation is

$$r_s = 1 - \left(\frac{6 \sum_i d_i^2}{N^3 - N} \right) \qquad \text{7.4}$$

Tables have been prepared which directly assess the significance of r_s when N is small. More generally, however, r_s is converted into a t value where

$$t = r_s \sqrt{\frac{N-2}{1-r_s^2}} \qquad 7.5$$

The distribution of t is closely approximated by Student's t distribution with $N-2$ degrees of freedom, hence the significance of t can be assessed by comparing t to the critical value, t_α, given in Table E, page 257. When $t \geqslant t_\alpha$ H_0 is rejected. (The t distribution is discussed in Chapter 9.)

From time to time one encounters ranked data in which 2 or more observations tie for a given rank. It is proper to apply a correction factor when tied rankings occur, but from a practical standpoint, the extra computations required by the correction are often not justified. For instance, Siegel (1956, pp. 207–10) presents an example with 3 pairs of ties amongst 12 observations: using equation 7.4, $r_s = \cdot617$ while the corrected formula yields a value of $\cdot616$ (the difference is very small indeed). Use of the correction is recommended under the following circumstances:

1. When three or more observations are tied equal.
2. When the number of pairs of ties is more than one quarter of the number of observations.

Because these two circumstances are not encountered very frequently, the version of Spearman's coefficient which corrects for ties will not be presented: it may be found in Siegel (1956, pp. 206–10).

Example

Like many Central American states, the republic of El Salvador is dominated by its capital city, San Salvador, in which more than one-tenth of the population of the country is concentrated. San Salvador is the source of most modernizing impulses in a country that is still largely agricultural and in which, in 1961, 52% of the population aged ten years or more were illiterate. Some progress has been made in the last decade, but levels of education are still rather low.

The problem to be examined is the relationship between accessibility to the capital city and levels of educational attainment in 1970. The two variables, which are measured for the fourteen departments of El Salvador (see Figure 7.1), are as follows:

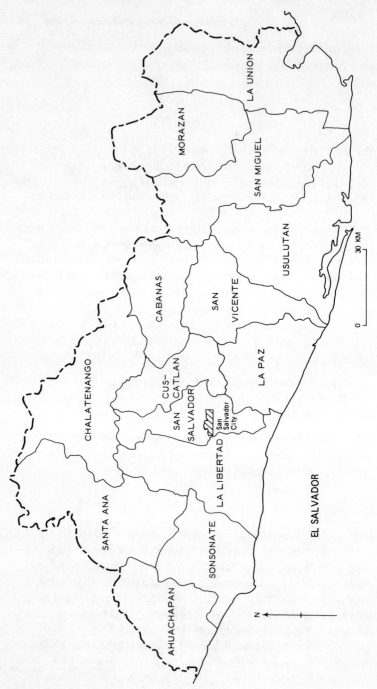

Figure 7.1: The departments of El Salvador.

X = distance from the population centroid of a department to San Salvador

Y = proportion of the population who have completed primary education.

The observations on both variables are ranked from 1 to 14, and in the case of Y, two departments, Cuscatlan and San Vicente, are tied equal (see Table 7.3).

Table 7.3: Education levels and accessibility to the city of San Salvador in the Republic of El Salvador, 1970.

Department	Ranked distance from centroid of department to San Salvador (X)	Ranked proportion of population who have completed primary education (Y)	d	d^2
Ahuachapan	5	13	−8	64
Santa Ana	8	7	1	1
Sonsonate	7	8	−1	1
Chalatenango	9	11	−2	4
La Libertad	13	5	8	64
San Salvador	14	1	13	169
Cuscatlan	12	3·5=	8·5	72·25
La Paz	11	2	9	81
Cabanas	6	12	−6	36
San Vicente	10	3·5=	6·5	42·25
Usulutan	4	6	−2	4
San Miguel	3	9	−6	36
Morazan	2	14	−12	144
La Union	1	10	−9	81
Total			0	799·5

SOURCE: *Annuario Estadistico 1970*, Volumen V, Direccion General de Estadistica Censos, Republica de El Salvador.

PROBLEM DEFINITION

1. H_0: the proportion of the population of a department who have completed their primary education is not correlated with the distance of a department from the city of San Salvador.

H_1: educational achievement is negatively correlated with distance from San Salvador.

The test is one-tailed.

2. α = ·01.

TEST SELECTION

1. Although initially measured at the interval scale, both X and Y were subsequently reduced to rank scale measurements.

2. $N = 14$.

3. Spearman's coefficient of rank correlation will be computed. The single tie present in Y will have very little influence on the value of r_s, hence the correction for tied rankings will not be used.

SAMPLING DISTRIBUTION

r_s will be tested for significance using the t distribution. In this case the test is one-tailed and

$$\nu = 14 - 2 = 12$$

Consulting Table E, page 257, with 12 degrees of freedom $t_{.01} = 2{\cdot}68$, hence if t is less than or equal to $-2{\cdot}68$, H_0 will be rejected (a negative sign is interpolated because a negative correlation is posited in H_1).

COMPUTATION

By arranging the data as in Table 7.3, the calculations required to obtain r_s are much simplified: Σd_i^2 is simply the total of the final column in this table which also provides a check on arithmetic errors (Σd_i ought to equal zero).

Evaluating **7.4**,

$$r_s = 1 - \left(\frac{6 \times 799{\cdot}5}{14^3 - 14} \right)$$

$$= 1 - \left(\frac{4797}{2730} \right)$$

$$= 1 - 1{\cdot}757 = -{\cdot}757$$

To obtain the t value using **7.5**,

$$t = -{\cdot}757 \sqrt{\frac{14 - 2}{(1 - (-{\cdot}757)^2)}}$$

$$= -{\cdot}757 \sqrt{\frac{12}{{\cdot}427}}$$

$$= -4{\cdot}01$$

which is a larger negative value than the critical value of t. H_0 is therefore rejected: the evidence indicates that the greater the distance between the centroid of a department and San Salvador, the lower are levels of educational achievement.

Geographical applications

In Chapter 1 it was noted that parametric tests are generally more powerful than the equivalent nonparametric tests. In the case of correlation coefficients, r_s is only 91% as efficient as Pearson's r in rejecting H_0, which might lead one to conclude that geographers ought to have little use for the rank correlation coefficient. While the Pearsonian coefficient has indeed been used by geographers much more frequently than r_s, there are three circumstances where the use of r_s is appropriate in geographical analysis.

First, there are situations where data are measured at the rank scale, so that a rank correlation coefficient is mandatory. A good example of this is to be found in the series of papers by Gould and his students on space preferences. For instance, in a study of perceptions in Western Nigeria, Gould and Ola (1970) asked participants to rank fifty hexagonal areas in terms of their attractiveness as a place of residence. Rank correlation coefficients comparing the responses of participants were calculated, and the resulting matrices composed of r_s values were then generalized using multivariate methods.

Rank correlation coefficients are also useful in instances where the assumptions of the parametric test are not met. As will become apparent in a later chapter, Pearson's r has the requirement that a sample be drawn from a population with a bivariate normal distribution. When the form of the population is clearly non-normal and cannot be normalized by data transformation, or when the form is not known and the sample is too small to give a clear picture of the population distribution, then use of a rank correlation coefficient is appropriate. A case in point is Amemiya's (1964) study comparing the economic diversity of the fifty-six largest Standard Metropolitan Areas in the United States in 1950 with their population ($r_s = \cdot 09$). Both of these variables tend to have non-normal distributions: the former often has a U-shaped arc-sine distribution, the latter a truncated exponential or log-normal distribution. The transformations required to normalize such distributions can be quite awkward, hence ranking the data provides a highly expedient solution.

The third circumstance in which geographers have found rank

correlation to be useful is when computational simplicity is an
important consideration. This applies particularly in places lacking
access to computers or electronic calculators. Although it is not made
explicit, this consideration may well have influenced Zaidi's choice of
r_s in a study published in 1968. Zaidi's data matrix consisted of nine
socio-economic variables measured over forty-five central places in
West Pakistan: he computed forty-five r_s values using data from this
matrix. Computing Pearson's r manually in this type of situation is a
daunting task, whereas computing r_s is much less arduous.

The point biserial coefficient of correlation

Statistical applications

Suppose that there are two variables, X and Y, to be correlated, and the
observations on Y are measured on a continuous scale, while observations on X are made on a dichotomous scale (i.e. values are either 0
or 1). In this situation, correlation between X and Y is measured using
a biserial coefficient of correlation. There are two such coefficients:

1. The Biserial Coefficient of Correlation (r_{bis}) devised by Karl
Pearson.
2. The Point Biserial Coefficient of Correlation (r_{pb}).

The relative merits of these two coefficients are discussed by Walker and
Lev (1953, pp. 261–71) who suggest that the latter is the more useful
for the following reasons. First, r_{bis} assumes that the dichotomous
variable (X) is continuous, whereas r_{pb} does not make assumptions
about the distribution of X. Second, r_{pb} has a range −1 to +1, whereas
for r_{bis} the range is unlimited. Third, the sampling distribution of the
point biserial coefficient is known and tests of significance are straightforward: the exact sampling distribution of r_{bis} is unknown. For these
three reasons the point biserial coefficient is presented here.

Point biserial correlation is interesting in that elements of both
parametric and nonparametric correlation are involved. As the assumptions
(below) indicate, the method is certainly not distribution-free. On the
other hand the binary variable may be measured at the ordinal (presence/
absence) or even the nominal scale. The method is also closely related
to the difference of means tests that are discussed in Chapter 9.

Assumptions

1. The binary variable, X, may be used to define two sub-samples,

one for which $X = 0$, and the other for which $X = 1$. The values of Y within each of these sub-samples should be normally distributed and should have equal variances.

2. The two sub-samples defined by X should not be radically different in size, and the more equal they are in size, the more accurate is the test.

Test statistic

Let X be the dichotomous variable, and each observation on X will be scored either 0 or 1.

N_0 is the number of individuals classed with 0 on X.

N_1 is the number of individuals classed with 1 on X.

Hence $N = N_0 + N_1$ = the total number of observations.

Y is a continuous variable that is divided into two sub-groups by X. The means of these two sub-groups are denoted by $\overline{Y}_0$ and $\overline{Y}_1$ respectively where

$$\overline{Y}_0 = \frac{\Sigma Y_{0i}}{N_0} \qquad\qquad 7.6$$

$$\overline{Y}_1 = \frac{\Sigma Y_{1i}}{N_1} \qquad\qquad 7.7$$

The standard deviation of Y ($\hat{s}_Y$) is

$$\hat{s}_Y = \sqrt{\frac{N\Sigma Y^2 - (\Sigma Y)^2}{N(N-1)}} \qquad\qquad 7.8$$

When these quantities have been calculated, then r_{pb} can be obtained from

$$r_{pb} = \frac{\overline{Y}_1 - \overline{Y}_0}{\hat{s}_Y} \sqrt{\frac{N_1 N_0}{N(N-1)}} \qquad\qquad 7.9$$

As in the case of Spearman's r_s, this correlation coefficient is tested for significance against Student's t distribution with $N-2$ degrees of freedom. In this case,

$$t = r_{pb} \sqrt{\frac{N-2}{1-r_{pb}^2}} \qquad\qquad 7.10$$

When $|t| < t_\alpha$ the null hypothesis is accepted, and when $|t| \geqslant t_\alpha$, H_0 is rejected.

Table 7.4: Population and general store location in central places in Snohomish county, Washington, USA

Central place	General store (X)	Population	Log. Population (Y)	(Y^2)
Snohomish	0	3494	3·543	12·553
Edmonds	0	2996	3·477	12·090
Lake Stevens	1	2586	3·413	11·649
Marysville	0	2460	3·391	11·499
Arlington	0	1915	3·282	10·772
Monroe	0	1684	3·226	10·407
Lowell	0	1600	3·204	10·266
Darrington	0	974	2·989	8·934
Mukilteo	1	900	2·954	8·726
Sultan	0	850	2·929	8·579
Beverly Park	0	725	2·860	8·180
Stanwood	0	720	2·857	8·162
Maltby	1	700	2·845	8·094
Granite Falls	0	600	2·778	7·717
Alderwood Manor	0	600	2·778	7·717
Lynnwood	0	500	2·699	7·285
East Stanwood	0	390	2·591	6·713
Gold Bar	0	325	2·512	6·310
Warm Beach	1	314	2·497	6·235
Silvana	1	300	2·477	6·136
Startup	0	300	2·477	6·136
Florence	1	300	2·477	6·136
Index	1	220	2·342	5·485
Machias	1	200	2·301	5·295
Oso	1	200	2·301	5·295
Cathcart	1	175	2·243	5·031
Bryant	1	150	2·176	4·735
Cedarhome	1	100	2·000	4·000
Robe	0	50	1·699	2·887
Getchell	0	25	1·398	1·954
Trafton	1	25	1·398	1·954
Verlot	1	20	1·301	1·693
Silverton	1	15	1·176	1·383
Total			84·591	230·008

Example

One of the basic relationships upon which central place theory is constructed is the relationship between the presence/absence of a given economic activity in a town and the population of a town. Some activities are found only in large towns, some in smaller villages and hamlets, while other activities are found in central places of all sizes. This example will examine the relationship between the location of general stores and the population of central places, using data given by Berry and Garrison (1958) in their classic study of Snohomish county, Washington.

General stores tend to be found in smaller places: in larger towns, their role is usually replaced by more specialized outlets such as grocery and hardware stores, hence an inverse relationship is expected. The data to be analysed is given in Table 7.4, where for X, a 1 indicates the presence of a general store in a town and 0 indicates absence.

PROBLEM DEFINITION

1. H_0: the presence/absence of a general store is not associated with the population of a central place.

H_1: there is a negative correlation between general store location and the population of a central place.
The test is one-tailed.

2. $\alpha = .01$.

TEST SELECTION

1. The presence/absence of a store is dichotomous variable, while the population of towns is measured at the interval scale.

2. $N_0 = 18; N_1 = 15; N = 18 + 15 = 33$.

3. The data in its raw form does not meet the assumptions of the significance test for the point biserial correlation coefficient (the data in both sub-samples are skewed and non-normal). However, transforming the data by taking the logarithm of the population of each town made the distributions of the two sub-samples fairly normal, and their variances quite similar ($s_0^2 = 0.318; s_1^2 = 0.373$). The variances of the two sub-samples were then compared using a test that is described in Chapter 10, and they were found to be not significantly different from one another. The transformed data (Y) meets the requirements for the above test, and the sub-samples are fairly equal in size.

SAMPLING DISTRIBUTION

r_{pb} is tested for significance by conversion to a t value which is compared to Student's t distribution with $N - 2$ degrees of freedom. In this case

$$\nu = 33 - 2 = 31$$

and the critical value given in Table E, page 257, $t_{.01} = -2.45$ (the negative sign is interpolated because a negative correlation is predicted in H_1).

COMPUTATION

The computations for r_{pb} are facilitated if the data is set up as in Table 7.4. The means of the two sub-samples are as follows: for towns lacking a general store,

$$\bar{Y}_0 = \frac{50 \cdot 690}{18} = 2 \cdot 816$$

and for towns with a general store

$$\bar{Y}_1 = \frac{33 \cdot 901}{15} = 2 \cdot 260$$

The standard deviation of Y ($\hat{s}_Y$) is calculated using the column totals from Table 7.4.

$$\hat{s}_Y = \sqrt{\frac{33(230 \cdot 008) - (84 \cdot 591)^2}{(33)(32)}}$$

$$= \sqrt{\frac{434 \cdot 627}{1056}} = \cdot 6415$$

With this information, the correlation coefficient becomes

$$r_{pb} = \frac{2 \cdot 260 - 2 \cdot 816}{\cdot 6415} \sqrt{\frac{(15)(18)}{(33)(32)}}$$

$$= - \cdot 8667 \sqrt{\cdot 2556} = - \cdot 438$$

In order to test r_{pb} for significance, the t value must be calculated. Here

$$t = - \cdot 438 \sqrt{\frac{(33 - 2)}{(1 - (- \cdot 438)^2)}}$$

$$= - 2 \cdot 71$$

The calculated t value is more negative than the critical value of $-2 \cdot 45$; H_0 is therefore rejected and we conclude that there is a significant negative correlation between the population of a town and the presence/absence of a general store. As an aside, the discrepancy between the r_{pb} reported in the original Snohomish county study ($- \cdot 347$), and the somewhat more negative value reported here ($- \cdot 438$) is attributed to the fact that Berry and Garrison did not apply a log transformation to the population variable.

Geographical applications

The major geographical application of point biserial correlation relates
to the locational characteristics of economic activities, and it includes
studies of both manufacturing industry and of tertiary activities.
Stafford's (1966) study of manufacturing location in Illinois is an
example of the former. Stafford correlated the presence/absence of 138
three digit SIC categories with the population of the 102 counties in
Illinois. With some exceptions, he found the correlations to be
surprisingly low, which suggests that population is generally not a very
good predictor of industrial location: however, some activities such as
sugar, tobacco, fabrics, fur, leather goods and jewellery industries are
concentrated in the counties with large populations, while food, wood,
newspaper and concrete industries are found mainly in smaller counties.
The example, above, drawn from work by Berry and Garrison (1958)
is perhaps the best known application of point biserial correlation to
central place studies. Berry and Garrison found twelve of the central
place attributes that they examined to be significantly correlated with
the size of central places in Snohomish county. The general store
example was the only significant negative relationship, while strong
positive correlations were found for such functions as public libraries,
sewage systems, and state liquor stores.

8   Court's method for map comparison

Statistical applications

Spatial analysis is concerned with identifying regularities in the geographical distribution of phenomena. It includes the study of point and line patterns, of directional preferences, of surface configurations and of contiguity: it also includes the comparison of distribution maps, a topic which is considered in this chapter. Two related map comparison tests will be discussed. Technically, this chapter is an extension of the previous chapter, but in this case correlation is measured explicitly in a spatial context.

The origins of the map comparison tests can be traced back to Quenouille's book on *Associated Measurements* (1952) in which a measure known as the *coefficient of medial correlation* is presented. This measure requires individuals to be cross-classified on two variables. For instance, we might cross-classify a commuter's age with the length of his journey-to-work. The cross classification is reduced to a 2 x 2 table, with one category of each variable representing values below the median, the other, values above the median. Quenouille presents a procedure for testing an observed medial correlation coefficient, q, for significance. This is an exact test for determining whether a given sample correlation could have been drawn from a population with a medial correlation of zero at a specified level of significance. If the null hypothesis is rejected, then a relation between the two variables is assumed to exist. Somewhat different is the method outlined in Kruskal's review paper on 'Ordinal measures of association' (1958, p. 836). In contrast to Quenouille's test which compares a sample correlation to a theoretical distribution, Kruskal's method places confidence intervals around a computed correlation coefficient, and may be used to determine whether two sample correlation coefficients are significantly different from one another, and hence whether two samples are drawn from the same population.

The geographical version of these tests was developed by Arnold

Court (1970) and involves comparing two distributions mapped over the same area. Both of the distributions are classified in such a way that half of the map area belongs to the above-median category, and the other half to the below-median category. By superimposing one map over the other, one creates a composite map with four categories: areas below the median on X and on Y; areas above the median on X and below on Y; areas below the median on X and above on Y; and areas above the median on X and on Y.

Court's method may be applied in two circumstances. In the first case, the two variables are measured over the same set of areal units: that is to say, the spatial sampling frames of the two variables are identical. In this case either an isopleth or a choropleth map may be prepared such that approximately half of the map area is allocated to each category of each variable. A medial correlation coefficient based on areas is then computed. From the geographer's viewpoint, this approach has the advantage that association is based on area, rather than the non-areal criteria on which most other measures of correlation are based. The weakness of this approach is that it usually involves discarding a lot of information and then spending considerable time measuring map areas to come up with a fairly weak measure of spatial association. The second circumstance where the coefficient may be used is when the spatial sampling frames of the two variables do not correspond. In this case the test is uniquely useful because one does not have the pairs of observations required for conventional correlation measures. For instance one may wish to compare domestic water consumption measured over water supply districts with socio-economic variables measured over census tracts. Isopleth maps of each variable are prepared and then superimposed on each other to create a composite map with four categories: the areal extent of each of these categories is measured and the medial correlation calculated.

One of the key operational problems involves measuring irregular areas on the composite map. Three methods, given in order of increasing sophistication, are: lay a tracing of the composite map over a piece of graph paper marked with a fairly fine grid, and count the number of squares occupied by each of the four categories; use a planimeter; use a map digitizer. Whichever method is used, one requires a 2 x 2 table in which the values give the percentages of the composite map belonging to each of the four categories. It is useful to know that the coefficient, q, is not materially affected if a supposed medial isoline does not define two exactly equal areas.

Assumptions

1. There is a bivariate distribution involving two variables, say X and Y.

2. Observations are measured at the ordinal scale so that for each variable the study region can be divided into two approximately equal areas, one with observations above the median, the other with observations below the median. A choropleth map may be used, but in the normal case two isolines representing the median for each variable (weighted by area) are drawn on a composite map.

Test statistic

1. An exact test comparing q *to a population with a medial correlation of zero.*

Given a resemblance matrix as follows:

		Variable X	
		Below (−) median	Above (+) median
Variable Y	Below (−) median	$(--)\,a$	$(+-)b$
	Above (+) median	$(-+)\,c$	$(++)\,d$

then the coefficient of medial correlation, q, is

$$q = \frac{(a+d)-(b+c)}{a+b+c+d} \qquad\qquad 8.1$$

q displays the two properties desired of correlation coefficients: it has a range +1 to −1, and a value of 0 indicates that the two variables are independently distributed. Numerical values of q should not, however, be equated with the numerical values of other correlation coefficients such as Spearman's r_s, although there is a relationship in the bivariate normal case (Kruskal, 1958).

q is a sample parameter, and like most sample parameters it may be used to make inferences about the population from which the sample is drawn. As Greek letters are generally used to represent population parameters, we will use the old Greek letter Ϙ (koppa) as a symbol for the population medial correlation coefficient. The basic test involves using q to determine whether $Ϙ \neq 0$. Clearly, when $Ϙ \neq 0$, we conclude that there is a significant statistical dependence between X and Y.

The significance of q is determined by computing a critical value, q_α where α is the chosen level of significance. Using N to denote the number of pairs of observations, then following Quenouille (1952)

$$q_\alpha = \frac{2}{N} + \frac{z}{\sqrt{N}} \qquad\qquad 8.2$$

where z is the z value (of a normal distribution) that is associated with the chosen level of α in a two tailed test. For the ·05 level of significance, $z = 1·96$; for the ·01 level of significance, $z = 2·58$. If the sampling frames do not coincide, then N is defined as the smaller number of areal units. Suppose, for example, that $\alpha = ·01$ and $N = 30$: then

$$q_{·01} = \frac{2}{30} + \frac{2·58}{\sqrt{30}} = ·538$$

When q is more positive or more negative than q_α, the null hypothesis is rejected. (Quenouille also gives a table of critical values of q, but formula **8.2** is very easy to evaluate, hence there is no need to reproduce the table.)

2. An approximate test to determine whether two values of q *are significantly different from one another.*

In this case there are at least three variables, let us say X, Y and Z. Taking two variables at a time, a resemblance matrix is tabulated and q is computed. One then makes pairwise comparisons of the medial correlation coefficients to determine whether they are significantly different from each other. In instances where a significant difference is established, one may conclude that the two sample coefficients are drawn from different populations.

The estimated standard error for a medial correlation coefficient is given by

$$s_q = \sqrt{\frac{(1-q)^2}{N}} \qquad\qquad 8.3$$

The difference between two sample q values, say q_{XY} and q_{XZ} is given by

$$t = \frac{q_{XY} - q_{XZ}}{\sqrt{s^2_{q_{XY}} + s^2_{q_{XZ}}}}$$ 8.4

and the distribution of t approximates Student's t distribution with $N-2$ degrees of freedom (Court, 1970).

Example

The sub-continent of India epitomizes many of the more difficult problems of economic development. For instance, in recent years increases in Gross National Product have largely been matched by increases in population so that real income per caput has virtually stood still. The census atlas of India for 1961 (in itself an excellent piece of work) contains plate after plate mapping various aspects of India's demise.

This example is concerned with the four Southern provinces of Andhra Pradesh, Mysore, Madras and Kerala. Three variables are considered: Population Change 1951–61 (X); Literacy Levels in 1961 (Y); and the Proportion of the Workforce aged fifteen to fifty-nine in 1961 who are cultivators (Z). Although these variables are measured over identical sampling frames consisting of 62 districts, for purposes of illustration the example will be developed as if the sampling frames were not coincident.

First test

The first test is to determine whether variables X and Y are spatially associated. Figure 8.1 is a composite of two isopleth maps showing the distribution of population change from 1951–61 and the distribution of levels of literacy in 1961. The percentage of the composite map area classed a, b, c and d was measured using a polar planimeter and the results are given in Table 8.1.

PROBLEM DEFINITION

1. H_0: the distributions of population change 1951–61, and literacy levels in 1961, are independent of one another (i.e. $\rho = 0$).

H_1: the distributions of these two variables are related, when the coefficient of medial correlation is used as a measure of association. The test is two-tailed.

2. $\alpha = \cdot 05$.

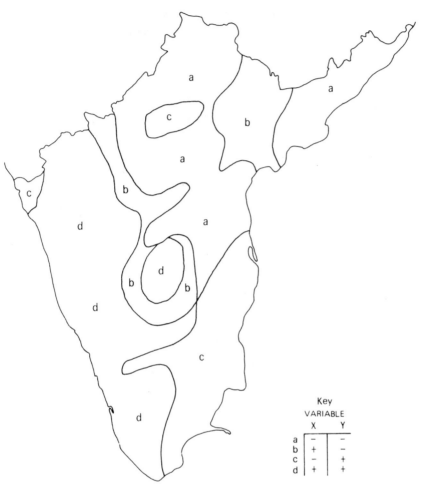

Figure 8.1: Composite map of population change 1951–61 (X) and literacy in 1961 (Y) for Southern India.

TEST SELECTION

1. Both *X* and *Y* are variables measured at the ordinal scale, and are separated into two categories about the median (weighted by area).

2. There are 62 districts hence *N* = 62.

3. The assumptions for Court's test are met: there is a bivariate distribution, measured at the ordinal scale.

Table 8.1:   Population growth 1951−61 and literacy in 1961 in Southern India (values give the percentage of the total area in each category)

		Population change 1951−61 (X)		
		Below (−) median	Above (+) median	*Total*
Literacy levels in 1961 (Y)	Below (−) median	33·5	15·7	49·2
	Above (+) median	17·8	33·0	50·8
	Total	51·3	48·7	100

SOURCE: Based on Plates 19 and 155, Census of India 1961, Volume 1, Part IX, Census Atlas.

SAMPLING DISTRIBUTION

The critical value of q is determined from **8.2**

$$q_{.05} = \frac{2}{62} + \frac{1 \cdot 96}{\sqrt{62}} = \cdot 032 + \cdot 249 = \pm \cdot 281$$

COMPUTATION

Computation is based on the data in Table 8.1. Using equation **8.1**, we have

$$q = \frac{(33 \cdot 5 + 33 \cdot 0) - (15 \cdot 7 + 17 \cdot 8)}{33 \cdot 5 + 15 \cdot 7 + 17 \cdot 8 + 33 \cdot 0}$$

$$= \frac{66 \cdot 5 - 33 \cdot 5}{100} = + \cdot 330$$

In this case q is somewhat larger than the critical value of $\pm \cdot 281$ and we can reject H_0. We conclude that $\rho \neq 0$ and that there is a significant positive spatial association so that, for instance, areas with an above median birth rate tend to coincide with areas with an above median literacy level. It would require considerable local knowledge to interpret this finding but the picture that emerges is not very encouraging from a developmental viewpoint. One might have hoped that birth rates would be lower in areas where people are more literate but in reality the reverse appears to be the case.

Second test

The second test involves comparing two q values, q_{XY} and q_{XZ} to determine whether they belong to the same population. This is an approximate test which involves placing confidence intervals around the sample parameters. q_{XY} was calculated in the first test and was found to be $+\cdot330$. q_{XZ} relates population growth 1951–61 to the percentage of the population who are cultivators: as Figure 8.2 shows, the spatial association is not very strong and the calculations based on the data in Table 8.2 show that $q_{XZ} = +\cdot080$.

Table 8.2: Population growth 1951–61 and the proportion of the workforce aged 15–59 who were cultivators in Southern India, 1961 (values give the percentage of the total area in each category)

| | | Population change 1951–61 (X) | | |
		Below median (−)	Above median (+)	Total
Proportion of the workforce aged 15–59 who are cultivators (Z)	Below median (−)	28·0	22·7	50·7
	Above median (+)	23·3	26·0	49·3
		51·3	48·7	100

SOURCE: Based on Plates 19 and 91, Census of India 1961, Volume 1, Part IX, Census Atlas.

PROBLEM DEFINITION

1. H_0: the two medial correlation coefficients q_{XY} and q_{XZ} do not differ significantly from one another and hence are drawn from the same population.

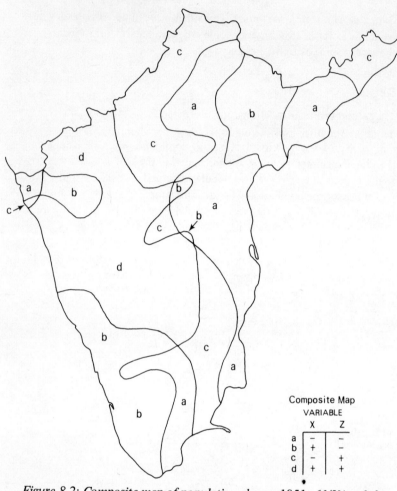

Composite Map
VARIABLE

	X	Z
a	–	–
b	+	–
c	–	+
d	+	+

Figure 8.2: Composite map of population change 1951–61(X) and the proportion of the workforce who were cultivators in 1961 (Z) for Southern India.

H_1: q_{XY} and q_{XZ} are significantly different and therefore are sample parameters drawn from different populations.

The test is two-tailed.

2. $\alpha = \cdot05$.

TEST SELECTION

The same conditions prevail as in the first test.

SAMPLING DISTRIBUTION

The distribution of the test statistic, t, is approximated by Student's t with $N-2$ degrees of freedom. As $N = 62$, $\nu = 60$ and consulting Table E, page 257, $t_{.05} = 2.00$.

COMPUTATION

The first step is to compute the variances of the two sample parameters.

$$s_{q_{XY}}^2 = \frac{(1 - \cdot330)^2}{62} = \frac{\cdot4489}{62} = \cdot00724$$

$$s_{q_{XZ}}^2 = \frac{(1 - \cdot080)^2}{62} = \frac{\cdot8464}{62} = \cdot01365$$

Using **8.4** to obtain t:

$$t = \frac{\cdot330 - \cdot080}{\sqrt{\cdot00724 + \cdot01365}} = \frac{\cdot250}{\cdot14445} = 1\cdot73$$

The computed value of t is less than the critical value of $2\cdot00$ hence we must accept the null hypothesis and conclude that q_{XY} and q_{XZ} are not significantly different from one another.

Geographical applications

In the paper in which he proposed this method, Court (1970) developed an example comparing the distribution of manufacturing employment in the northwest United States with the distribution of high incomes and of low incomes: however, Court's concern was with methodology and he does not comment on the substantive meaning of his results. Apart from this example there is a lack of published applications of Court's method due both to the recency of its development and to a lack of appreciation of its value, especially in the case of mis-matched sampling frames. For instance, there is an interesting potential application in electoral geography: voting behaviour, which is recorded for voting precincts, could be correlated with socio-economic data, which are recorded for census tracts (the boundaries of voting precincts and census tracts only rarely coincide). Despite being based on a fairly insensitive measure of correlation, Court's method is presented for two reasons: first, it is uniquely useful in comparing map distributions with mis-matched sampling frames; and second, it provides the simplest approach to measuring correspondence between spatial distributions — a topic that will be taken up again in the final chapter.

Further readings in nonparametric statistics

Amongst the burgeoning literature on nonparametric methods, the following provide useful introductions to topics not covered in this book: Conover (1971), Hajek (1969), Pierce (1970), Siegel (1956).

Part three
Parametric methods

9 Analysis of means

Four 'difference of means' tests are discussed in this chapter, as follows:

1. Comparing a sample mean to a population mean when the population standard deviation is known.
2. Comparing a sample mean to a population mean when the population standard deviation is not known.
3. Comparing the means of two independent samples with equal variances.
4. Comparing the means of two independent samples with unequal variances when, for both samples, $N \geqslant 25$.

Examples will be presented for all but the second case which is sufficiently similar to the first case that a separate example is not warranted. Geographical applications will be discussed at the end of the chapter and not after each test. Tests that are used to compare the means of more than two samples are dealt with in Chapter 10.

Three of the four tests make use of Student's t distribution. Since only passing reference has been made to this distribution in previous chapters, some background information is warranted. This distribution was discovered by W. C. Gosset, a remarkable Irishman who for many years was an employee of the Guinness Brewing Company. This company had a policy that employees publishing papers resulting from their work for the company should maintain anonymity: accordingly Gosset published under the pseudonym 'Student', hence Student's t distribution.

Like the normal distribution, the t distribution is symmetrical, but whereas the former has a uniform shape regardless of sample size, the t distribution changes shape according to the number of degrees of freedom. When there are only a few degrees of freedom, the t distribution is flatter and more spread out than the normal curve (i.e. it is platykurtic). As the number of degrees of freedom increases, so the t distribution more closely approximates the normal curve until, in the limiting case, they become identical (see Figure 9.1). It is often suggested in older textbooks that when $N \geqslant 30$, the normal approximation to the t distribution should be used. There is a historical reason for this, namely that for many years the *t* distribution was tabled only for $v \leqslant 30$. Tables for $v > 30$ are now available, hence in situations where Student's *t* is the appropriate sampling distribution, it should be used in preference to the normal distribution for samples of any size.

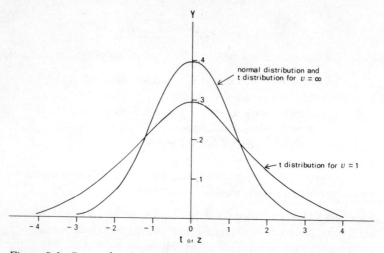

Figure 9.1: Curves for the t *distribution and the normal distribution.*

Comparing a sample mean to a population mean when the population standard deviation is known

Statistical applications

Given a sample mean, $\bar{X}$, a population mean, μ (mu), and a population standard deviation, σ (sigma), this test is used to determine the probability that the mean value of a given sample could have been drawn randomly from a population with known mean and variance: symbolically, we are asking whether $\bar{X} = \mu$.

Assumptions

1. For small samples with $N < 30$, the population should be normally distributed. For large samples this assumption is not required provided that the population is at least twice as large as the sample.

2. The population mean and standard deviation are known.

3. The sample is drawn randomly from the population.

Test statistic

Before presenting the test statistic, the concept of the sampling distribution of the means will be explained using a simple example. Suppose there is a population consisting of five observations (3, 5, 6, 7, 9) and we are drawing samples consisting of two observations, in each case calculating the mean of the two values. In this example, the sampling distribution of the means based on all possible pairs of observations, is as follows:

$(3 + 5)/2 = 4$	$(5 + 7)/2 = 6$
$(3 + 6)/2 = 4{\cdot}5$	$(5 + 9)/2 = 7$
$(3 + 7)/2 = 5$	$(6 + 7)/2 = 6{\cdot}5$
$(3 + 9)/2 = 6$	$(6 + 9)/2 = 7{\cdot}5$
$(5 + 6)/2 = 5{\cdot}5$	$(7 + 9)/2 = 8$

It should be clear that the sampling distribution of the means describes the distribution of the mean values of all samples of size N drawn randomly from a population.

Fortunately we do not normally have to evaluate the sampling distribution of the means in this laborious manner because, under the above assumptions, the sampling distribution of the means is known to be normally distributed, and the standard deviation of the sampling distribution of the means is given by

$$\sigma_{\overline{X}} = \frac{\sigma}{\sqrt{N}} \qquad \textbf{9.1}$$

Having computed $\sigma_{\overline{X}}$, one then proceeds to express the sample mean in standard units about the population mean. Thus

$$z = \frac{\overline{X} - \mu}{\sigma_{\overline{X}}} \qquad \textbf{9.2}$$

and the probability associated with any given z value can be looked up in Table F, page 258.

9.1 is used both with samples drawn from an infinite population, and with samples drawn *with replacement* from a finite population (replacing an individual so that it can be drawn again has the effect of making a population infinite). When a sample is drawn *without replacement* from a finite population, then a correction should be applied to **9.1** as follows:

$$\sigma_{\overline{X}} = \frac{\sigma}{\sqrt{N}} \sqrt{\frac{N_p - N}{N_p - 1}} \qquad \textbf{9.3}$$

where N_p is the total population.

Example

Prince Edward Island, which lies in the Gulf of St Lawrence, is Canada's smallest province. The island's economy is dominated by agriculture, indeed it is somewhat inaccurately referred to as 'the million-acre farm' in some textbooks. Over the years there has developed a relatively prosperous farming economy in the central part of the province, particularly around Charlottetown and Kensington, and a moderately successful farming economy in Prince County in the western end of the island. However, to the east of the Hillsborough River in Kings county and parts of Queens county, many farms are quite marginal enterprises: indeed there has been considerable abandonment of land in this eastern region in recent years.

Prince Edward Island's population in 1961 was fairly widely dispersed through a network of 86 small urban centres mainly serving the farming community. Figure 9.2 maps the distribution of these central places. Treating the 86 urban centres as the population, then in the population as a whole the mean distance (in miles) between nearest neighbouring centres in 1961 was 3·440 miles with a standard deviation of 1·193 miles. The problem in this example is to determine whether a sample of central places drawn from the eastern part of the island has a significantly greater mean spacing than in the population as a whole; this might occur because of the region's low income generating capacity. Alternatively, the spacing might remain constant, but with fewer establishments and functions present in each central place. The average distance between nearest neighbours for the 27 central places in the eastern end of the province was 4·093 miles.

PROBLEM DEFINITION

1. H_0: the average spacing of central places in the eastern part of Prince Edward Island is not greater than the average spacing in the island as a whole.

H_1: in the eastern part of Prince Edward Island, central places are spaced further apart than in the Island as a whole.
The test is one-tailed.

2. $\alpha = \cdot05$

TEST SELECTION

1. Distances separating nearest neighbouring central places are measured at the interval scale.

Figure 9.2: Central places in Prince Edward Island, 1961.

2. $N = 27$. As sampling has taken place without replacement, the population size is required: $N_p = 86$.

3. We have a population consisting of the distances separating the 86 central places in the province, and a sample comprising 27 centres drawn from the eastern end of the island. As $N < 30$, the population should be normally distributed.

A histogram plotted for the population showed a curve with slight positive skewness, but generally approximating the normal form quite closely. For the population, μ and σ are known and for the sample $\bar{X}$ is known, hence it is appropriate to employ a difference of means test comparing a sample to a population in which σ is known.

The assumption which raises a problem both here and in many geographical examples concerns the randomness of the sample. It can be argued that the sample is random in the sense that stochastic processes influence the existence or non-existence of a central place. This is reflected in the disappearance of central places as they lose all their functions and the appearance of new central places as settlements take on a formal service function. From this perspective the sample can be viewed as belonging to a hypothetical universe of central place distributions in eastern Prince Edward Island. Needless to say, those who disavow the random space economy would eschew this interpretation.

SAMPLING DISTRIBUTION

In this case $\sigma_{\bar{X}}$ is normally distributed. As we are conducting a one-tailed test and $\alpha = \cdot01$, consulting Table F, page 258, we find $z_{.01} = 2 \cdot 33$. H_0 will be rejected if $z > z_{.01}$.

COMPUTATION

The key parameter to be calculated is the standard deviation of the sampling distribution of the means, as given in **9.3**: the correction is applied because sampling took place without replacement.

$$\sigma_{\bar{X}} = \frac{1 \cdot 193}{\sqrt{27}} \sqrt{\frac{86 - 27}{86 - 1}} = \cdot2296 \times \sqrt{\cdot6941} = \cdot1913$$

Computing the z value as in 9·2,

$$z = \frac{4 \cdot 093 - 3 \cdot 440}{\cdot1913} = 4 \cdot 41.$$

z is considerably larger than the critical value hence H_0 is rejected. It would appear that the distance separating nearest neighbouring central places in eastern Prince Edward Island is significantly greater than in the population as a whole.

Comparing a sample mean to a population mean when the population standard deviation is not known

Statistical applications

This test differs from the previous one in two respects: σ is not known and has to be estimated from sample data; instead of the normal distribution, Student's t is the appropriate sampling distribution. In other respects, this test is very similar to the preceding one.

Assumptions

They are identical to those of the previous test except that σ has to be estimated from the sample data.

Test statistic

In this case **9.1** may not be used because σ is unknown. However in Chapter 3 it was explained how sample parameters can be used to estimate population parameters: in particular **3.21** gave a formula for estimating σ from sample data. The estimate was:

$$\hat{s} = \sqrt{\frac{\Sigma X^2}{N-1} - \frac{(\Sigma X)^2}{N(N-1)}} \qquad \textbf{3.21}$$

Inserting $\hat{s}$ into 9·1 we obtain

$$\hat{s}_{\overline{X}} = \frac{\hat{s}}{\sqrt{N}} \qquad \textbf{9.4}$$

The test statistic then becomes

$$t = \frac{\overline{X} - \mu}{\hat{s}_{\overline{X}}} \qquad \textbf{9.5}$$

and this is compared to the appropriate t value in Table E, page 257, with $N-1$ degrees of freedom.

Comparing the means of two independent samples with equal variances

Statistical applications

The purpose of this test is to compare the means of two independent samples. It is assumed that the population parameters are unknown, but that sample means and variances are known: the sample variances are assumed to be equal. Sample data is then used to estimate the mean of the population from which each sample is drawn in order to answer the question, does $\mu_1 = \mu_2$?

Assumptions

1. For small samples with $N < 30$, the samples should be normally distributed, but for large samples this assumption may be relaxed.

2. The sample variances are drawn from the same population so that $\hat{s}_1^2 = \hat{s}_2^2$ (subject to sampling error).

3. There are two independent random samples.

Test statistic

Since the population variance is unknown, the sample variances are pooled to give an estimate of σ. The pooling procedure involves weighting each sample variance ($\hat{s}_1^2$ or $\hat{s}_2^2$) in accordance with the respective sample size (N_1 or N_2), which is logical since the larger a sample, the more reliable the estimate of σ. The pooled estimate of the variance is

$$\hat{s}^2 = \frac{\hat{s}_1^2 (N_1 - 1) + \hat{s}_2^2 (N_2 - 1)}{N_1 + N_2 - 2} \qquad \textbf{9.6}$$

This variance estimate is used to estimate the standard deviation of the sampling distribution of the means according to the formula

$$\hat{s}_{X_1 - X_2} = \sqrt{\frac{\hat{s}^2 (N_1 + N_2)}{N_1 N_2}} \qquad \textbf{9.7}$$

A t value may then be computed from

$$t = \frac{\bar{X}_1 - \bar{X}_2}{\hat{s}_{X_1 - X_2}} \qquad \textbf{9.8}$$

and this is compared to Student's t with $N_1 + N_2 - 2$ degrees of freedom. When $|t| < t_\alpha$, H_0 is accepted.

Example

Since the late 1960s, a technique in vogue amongst economic geographers is shift and share analysis. In essence, the purpose of this technique is to separate the change in a region's employment over a given time period into two components, change due to the regional mix

Table 9.1: Changes in employment, per 1000 employed persons in 1950, during the period 1950–60 that are attributed to the industrial mix of a state

Heartland (N_1 = 22)		Hinterland (N_2 = 27)	
Maine	− 23·05	Missouri	− 19·38
New Hampshire	− 1·97	North Dakota	−182·52
Vermont	− 54·63	South Dakota	−149·01
Massachusetts	+ 69·09	Nebraska	− 82·09
Rhode Island	+ 30·21	Kansas	− 44·37
Connecticut	+ 94·10	Tennessee	− 70·00
New York	+ 71·65	North Carolina	−134·55
Pennsylvania	+ 16·81	South Carolina	−152·11
New Jersey	+ 76·63	Georgia	− 96·10
Delaware	+ 30·01	Florida	+ 1·78
Maryland	+ 70·74	Alabama	−114·22
Virginia	+ 8·43	Mississippi	−188·74
West Virginia	−103·95	Louisiana	− 53·79
Kentucky	−109·84	Arkansas	−152·00
Illinois	+ 47·57	Oklahoma	− 51·89
Indiana	+ 11·20	Texas	− 18·56
Ohio	+ 40·43	New Mexico	− 18·40
Michigan	+ 1·25	Arizona	− 28·07
Wisconsin	− 26·05	Montana	−110·78
Minnesota	− 62·77	Idaho	−115·05
Iowa	− 69·15	Wyoming	− 78·82
District of Columbia	+130·66	Colorado	− 4·20
		Utah	− 27·09
		Washington	+ 60·23
		Oregon	− 41·80
		Nevada	− 29·86
		California	+ 69·11

SOURCES: Paraskevopoulos (1971) and 1950 United States Census of Population, Vol. II, Part I, Table 80.

of economic activity and a competitive element. For instance a region may experience a drop in employment levels because of a local concentration of declining heavy industries or agriculture; when the effect of this unfavourable mix of industries is taken into account, it may transpire that such a region has performed better than average.

This example is concerned with the change in employment due to the mix of economic activity in the forty-nine coterminous states of the United States (including Washington DC) for the period 1950–60. The states are split into two groups: the 'heartland' group coincides roughly with the manufacturing belt in the north east, while the 'hinterland' consists of the rest of the country. Table 9.1 records the change in employment due to the mix of industry per 1000 employees in 1950. Thus in Maine, for every 1000 employees in the base year, just over 23 jobs were lost in the following decade as a result of the somewhat unfavourable mix of industry in that state.

It is hypothesized that slow growing resource-oriented industries predominated in the hinterland in the 1950s, whereas in the heartland faster growing high-value-added industries were strongly represented. The average growth rate attributed to industrial composition during the 1950s in the 22 heartland states was +11·24 employees per 1000, whereas the average rate in the 27 hinterland states was −67·86 per thousand: could this difference have occurred by chance?

PROBLEM DEFINITION

1. H_0: The mean change in employment (per 1000 people) from 1950 to 1960 attributable to the mix of industry did not differ between the heartland states and the hinterland states of the United States.

H_1: The mean change in employment from 1950–60 attributable to the mix of industry was greater in the heartland than in the hinterland. The test is one-tailed.

2. $\alpha = ·01$

TEST SELECTION

1. Rates of change in employment are measured at the interval scale.
2. $N_1 = 22$ and $N_2 = 27$.
3. For $N < 30$, the samples are assumed to be drawn from normally distributed populations: Figure 9.3 plotting the histograms for the two samples, shows their distributions to be roughly normal. The two sample

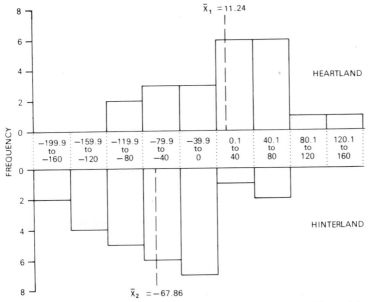

Figure 9.3: Histogram of employment change rates (per 1000 people) in the United States 1950–60 (see text).

variances are quite similar: $\hat{s}_1^2 = 4090.98$ while $\hat{s}_2^2 = 4537.80$ (a formal comparison of these two variances will be made in Chapter 10).

As in the previous example, the problematical assumptions are those of independence and randomness. The situation is non-experimental so that one cannot guarantee independence. The key question is whether the amount of employment in a given activity in a state is entirely governed by factors within the state, or whether it is influenced by the level of activity in that industry in neighbouring states. To take an unequivocal case, is the absence of skiing in Florida due to the absence of skiing in nearby Louisiana, or is it due to the sub-tropical climate of Florida? Obviously, the latter explanation is the key. Unfortunately, in the manufacturing sector inter-industry dependencies make it difficult to rule out the influence of neighbouring states, although the fairly large size of most states does minimize this problem. The question of independence is a thorny one that is not easily resolved: it is raised here because geographers often encounter the problem. Concerning the randomness assumption, one again has to view employment mix in 1950 as the outcome of a stochastic process if this assumption is to be met.

SAMPLING DISTRIBUTION

The test statistic in this example is $t_{.01}$ (for a one-tailed test) with $N_1 + N_2 - 2$ degrees of freedom. $v = 22 + 27 - 2 = 47$; consulting Table E, page 257, $t_{.01} = 2.41$.

COMPUTATION

The two sample variances must first be pooled to obtain an estimate of the population variance according to **9.6**.

$$\hat{s}^2 = \frac{(4090.98 \times 21) + (4537.80 \times 26)}{47} = 4338.16$$

Applying this parameter to **9.7** we obtain the standard deviation of the sampling distribution of the means.

$$\hat{s}_{X_1 - X_2} = \sqrt{\frac{4338.16(22 + 27)}{22 \times 27}} = \sqrt{357.86} = 18.92$$

From **9.8** we obtain the t value:

$$t = \frac{11.24 - (-67.86)}{18.92} = \frac{79.10}{18.92} = 4.18$$

Since t is considerably larger than $t_{.01}$, we reject H_0, concluding that the rate of employment change attributable to industrial mix was higher in those states in the heartland than in the hinterland from 1950 to 60.

Comparing the means of two independent samples with unequal variances

Statistical applications

This test differs from the previous test in that variances of the populations are not assumed to be equal. This gives rise to a test that is less powerful than the previous one.

Assumptions

1. Both N_1 and $N_2 \geqslant 25$: this test should not be applied to small samples.
2. There are two independent random samples.

Test statistic

Since it is assumed that $\sigma_1^2 \neq \sigma_2^2$, it is not possible to pool the two sample variances. The standard deviation of the sampling distribution of the means is therefore estimated from

$$\hat{s}_{\overline{X}_1 - \overline{X}_2} = \sqrt{\frac{\hat{s}_1^2}{N_1} + \frac{\hat{s}_2^2}{N_2}} \qquad 9.9$$

As in the previous test, $\hat{s}_{\overline{X}_1 - \overline{X}_2}$ is applied to 9.8 so that

$$t = \frac{\overline{X}_1 - \overline{X}_2}{\hat{s}_{\overline{X}_1 - \overline{X}_2}} \qquad 9.8$$

and this is compared to t with $N_1 + N_2 - 2$ degrees of freedom. H_0 is rejected when $|t| \geq t$.

Example

This example if taken from a study of perceptions in Kitchener–Waterloo in Ontario (Norcliffe, 1974). Kitchener and Waterloo are twin cities which in 1970 were governed by two separate municipal governments. However at that date there was afoot a plan to unite the cities under a single government – a plan which met strong resistance from the residents of Waterloo, the smaller of the two municipalities.

For both Kitchener and Waterloo, a random sample of interviewees were asked to rate both towns with respect to 23 bi-polar adjectives using a seven-point semantic scale. This example is based on the following bi-polar adjective:

Small variety of shops | _ _ _ _ _ _ $\underline{\times}$ | Great variety of shops

Each respondent was asked to place an X in one of the seven positions on the scale with respect to his or her image of Kitchener (or Waterloo). In this illustration an X was placed in the seventh position, which indicates that this hypothetical respondent perceives Kitchener to possess a great variety of shops.

68 interviews were completed in Kitchener (sample 1). The mean response ($\overline{X}_1$) was 5·65 with a standard deviation of 1·39. In Waterloo (sample 2) $N_2 = 45$, the mean is higher ($\overline{X}_2 = 6·09$), but the standard deviation is smaller ($\hat{s}_2 = 0·91$). The interview locations are plotted in Figure 9.4.

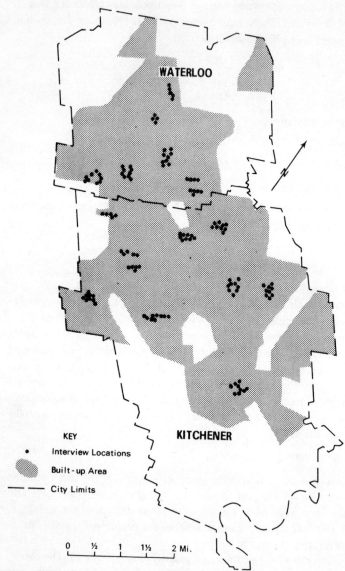

Figure 9.4: Interview locations in Kitchener–Waterloo.

PROBLEM DEFINITION

1. H_0: the mean response on the variety of shops in Kitchener is the same for respondents in both Kitchener and Waterloo.

H_1: citizens of Kitchener and of Waterloo differ in their average assessment of the variety of shops in Kitchener so that $\mu_1 \neq \mu_2$.
The test is two-tailed.

2. $\alpha = \cdot 05$.

TEST SELECTION

1. Strictly speaking, the data is measured at the ordinal scale since interviewees could only respond at integer positions on the scale. However, it is possible to view the responses as a continuous distribution generalized to integers. Furthermore standard deviations can be computed for this kind of data and as Anderson (1961) demonstrates tests based on Student's t are applicable.

2. $N_1 = 68$, $N_2 = 45$ hence for both samples $N > 25$.

3. There are two independent random samples. $\hat{s}_1^2 = 1 \cdot 932$ while $\hat{s}_2^2 = 0 \cdot 828$, hence the variances are quite different from each other; a difference of means test for two independent samples (with the assumption that $\sigma_1^2 \neq \sigma_2^2$) will be applied.

SAMPLING DISTRIBUTION

The critical value for this test is t with $N_1 + N_2 - 2$ degrees of freedom. Consulting Table E, page 257, for $\alpha = \cdot 05$ and $v = 111$ in a two-tailed test, we find that $t_{\cdot 05} = 1 \cdot 98$.

COMPUTATION

The standard deviation of the sampling distribution of the means according to 9.9 is

$$\hat{s}_{\overline{X}_1 - \overline{X}_2} = \sqrt{\frac{1 \cdot 932}{68} + \frac{0 \cdot 828}{45}} = 0 \cdot 22$$

The t value is then given by 9.8:

$$t = \frac{5 \cdot 65 - 6 \cdot 09}{0 \cdot 22} = \frac{- \cdot 44}{\cdot 22} = -2 \cdot 00$$

Since $|t| > t_\alpha$, the null hypothesis is rejected and we conclude that residents of Waterloo perceive there to be a significantly greater variety of shops in Kitchener than do the residents of Kitchener.

Geographical applications

Of the various difference of means tests discussed in this chapter, geographers have made far greater use of the tests comparing two sample means. One of the few published examples of a test comparing a sample mean to a theoretical population mean is that of Rushton, Golledge and Clark (1967) in which they compare such things as the mean length of shopping trips to mean values predicted by a normative model.

Two sample tests have been applied by geographers in two different circumstances; first, to the comparison of two regional means; and second, to compare the spatial attributes of two different populations which may or may not occupy the same region. Haggett (1965, p. 291) developed an example of the former in comparing the mean spacing of towns in East Texas, and in Iowa and Missouri using data drawn from Lösch. The example of perceptions in Kitchener–Waterloo was drawn from one of four sets of difference of means tests aimed at identifying the influence of urban territoriality on the images of residents of the twin cities (Norcliffe, 1974).

Being relatively simple and straightforward, difference of means tests are of considerable value to spatial analysts. Their major limitation is that they provide insights into differences of means – and only that: they do not provide insights into the more intriguing question of association. Although difference measures are used quite commonly, the major use that geographers have for inferential statistics is to identify associations between phenomena. It is this central interest in association that makes correlation methods the techniques most frequently applied in spatial analysis.

10   Analysis of variance

The term 'analysis of variance' is used in two different senses in this chapter. First, it is used in the conventional statistical sense to describe tests which generalize the difference of means tests discussed in the previous chapter. This generalization involves comparing the means of two or more samples and/or the effect of two or more sources of variation on a given phenomenon. Two tests of this type are presented in the first part of this chapter, namely one way and two way analysis of variance, the latter being a generalization of the former. The other sense in which the term is used is to compare the variances of two samples to determine whether they are significantly different from one another. This is the most useful of a number of tests that can be applied to standard deviations and variances (see Walker and Lev (1969, Chapter 15) and Croxton, Cowden and Klein (1968, Chapter 26)).

The F distribution, which is the required sampling distribution for all of the tests discussed in this chapter, was originally conceived by R. A. Fisher. Later, the distinguished Indian statistician, P. C. Mahalanobis, proposed the formal statistic, calling it F after Fisher. Since the F distribution is the most complicated sampling distribution encountered in this text, its use deserves some explanation. It has already been established that the normal distribution has a fixed shape regardless of sample size, whereas the shapes of the t and chi square distributions are governed by the number of degrees of freedom that are available. The shape of the F distribution is governed by *two* parameters, both involving degrees of freedom, the reason for this being that the F test is a 'variance ratio test'. In other words, the F test involves comparing two variances which become the numerator and denominator of an equation: there are ν_1 degrees of freedom for the numerator variance and ν_2 degrees of freedom for the denominator variance.

The table of F values in the Appendix of this book, pages 260–61, gives critical values of F only for $\alpha = \cdot01$ and $\alpha = \cdot05$. Suppose we wished to look up $F_{\cdot05}(5,2)$: $\nu_1 = 5$ and one looks for the column headed 5 at the top of Table H: $\nu_2 = 2$ and one enters the second row at the side of the table to find a critical value of 19·30. Following the same rules, $F_{.05}(24, 60) = 1·70$; $F_{.01}(24, 60) = 2·12$. Note that the table is asymmetric so that, for instance, $F_{.01}(3, 12) = 5·95$, but $F_{.01}(12, 3) = 27·05$.

It can be shown that the F distribution is a generalization of Student's t. Comparing the column of F for $\nu_1 = 1$ with t, one finds that the t values are the square root of the corresponding F values (in both cases assuming a two-tailed test). More formally,

$$t_\alpha (k) = \sqrt{F_\alpha (1, k)} \qquad\qquad 10.1$$

From this perspective t tests can be viewed as variance ratio tests in which there is always one degree of freedom associated with the numerator.

One way analysis of variance

Statistical applications

Analysis of variance forms a major topic in statistics, indeed several textbooks are devoted largely to this topic. In this introductory text there is not the space to deal with some of the more sophisticated variations, hence the discussion will concentrate on the most straightforward versions of one way, and two way analysis of variance. For a more advanced treatment, see Gaito (1973). A corresponding nonparametric test, the Kruskal–Wallis test, is summarized in Siegel (1956, Chapter 8).

Newcomers to statistics often find analysis of variance the most difficult of techniques to grasp. Conceptually, it is easy to understand: what appears to be off-putting are the equations with numerous summation signs and sub-scripts. However, once familiar with the conceptual basis, the equations should become much more meaningful. The one way case is dealt with first because it is the most straightforward.

It was stated in Chapter 3 that variance is the measure of dispersion most commonly used by statisticians. Analysis of variance involves separating total variance into components that are attributed to different sources. One way analysis of variance separates the total sample variance into two distinct components, one of which is identified with the chosen explanatory variable, the other with random effects. Estimates of variance derived from these two components are then expressed as an F ratio, and tested against the F distribution, to see whether the explanatory variable accounts for a significant proportion of the total variance.

Assumptions

1. For the basic model presented here, each category of each variable must have an equal number of observations, measured at the interval or ratio scale.

2. One way analysis of variance assumes a model in which the mean value in category j ($\bar{X}_j$) is the grand mean ($\bar{\bar{X}}$), plus (or minus) the effect due to the explanatory variable (Y_j), plus (or minus) an error term (e_j). Symbolically then

$$\bar{X}_j = \bar{\bar{X}} + Y_j + e_j \qquad \qquad \textbf{10.2}$$

In **10.2**, the error terms should:

(a) be normally distributed;
(b) have means of zero;
(c) have equal variances (i.e. be homoscedastic);
(d) be mutually uncorrelated (i.e. their covariance = 0).

These four requirements for the error terms might appear to be very demanding. Fortunately, the model is still valid even when these requirements are not rigidly met. Specifically, Cochran (1947) reports on several studies which indicate that no serious errors of inference are likely if the data are non-normal, and this has been corroborated by Donaldson (1968): besides, in cases of severe non-normality, transformations may be applied. Since non-normal error terms will probably have non-zero means, the second assumption may also be relaxed. Donaldson defines robustness more precisely with respect to Type I and Type II errors: in essence, he finds the test robust and most commonly conservative in cases of non-normal (exponential and lognormal) distributions, and of unequal variances. It would appear, then, that the assumptions concerning the error terms may be relaxed in testing against H_0, although as the violation of the assumptions becomes more marked, so the test loses some of its efficiency.

3. The sample data should be randomly drawn from the relevant universe.

Test statistic

Assume that one way analysis of variance is being applied to data as in Table 10.1, with Variable I having four categories, and three observations in each category. This table contains the basic quantities required for one way analysis of variance, namely the sum of the Xs and the Xs squared, and the grand totals for the whole table. Now,

Table 10.1: Tabular representation of one way analysis of variance

Category	Variable I A	B	C	D	Grand totals
	X_{11} X_{12} X_{13}	X_{21} X_{22} X_{23}	X_{31} X_{32} X_{33}	X_{41} X_{42} X_{43}	
Sum of Xs	$\sum\limits_{k} X_{1k}$	$\sum\limits_{k} X_{2k}$	$\sum\limits_{k} X_{3k}$	$\sum\limits_{k} X_{4k}$	$\sum\limits_{j}\sum\limits_{k} X_{jk}$
Sum of Xs squared	$\sum\limits_{k} X_{1k}^2$	$\sum\limits_{k} X_{2k}^2$	$\sum\limits_{k} X_{3k}^2$	$\sum\limits_{k} X_{4k}^2$	$\sum\limits_{j}\sum\limits_{k} X_{jk}^2$

let m be the number of categories ($j = 1, m$)

let n be the number of observations in each category ($k = 1, n$)

and let N be the grand total number of observations ($N = m \times n$).

Two sums of squares are now required, the total sum of the squares (SS_T), where

$$SS_T = \sum_j \sum_k X_{jk}^2 - \frac{\left(\sum\limits_j \sum\limits_k X_{jk}\right)^2}{N}$$ 10.3

and the sum of the squares due to the columns, SS_C, which represents the effect of Variable I

$$SS_C = \frac{\sum\limits_j \left(\sum\limits_k X_{jk}\right)^2}{m} - \frac{\left(\sum\limits_j \sum\limits_k X_{jk}\right)^2}{N}$$ 10.4

The error sum of the squares, SS_E, is obtained easily from

$$SS_E = SS_T - SS_C$$ 10.5

Having obtained these three sums of squares, testing is best done by setting up a table using the format in Table 10.2.

Table 10.2: Testing in one way analysis of variance

Source of variation	Sums of squares	Degrees of freedom	Estimate of variance	F
Total	SS_T	$N-1$		
Variable I	SS_C	$m-1 = \nu_1$	$SS_C \div \nu_1 = \hat{s}_C^2$	$\hat{s}_C^2 \div \hat{s}_E^2$
Error	SS_E	$N-m = \nu_2$	$SS_E \div \nu_2 = \hat{s}_E^2$	

The number of degrees of freedom associated with the numerator $\nu_1 = m - 1$, and with the denominator, $\nu_2 = N - m$. The estimates of variance for the columns (Variable I), and for the errors, are obtained by dividing the sums of the squares by the respective number of degrees of freedom. The two estimates of variance are then expressed as a ratio as in **10.6**.

$$F = \frac{\hat{s}_C^2}{\hat{s}_E^2} \qquad \qquad \textbf{10.6}$$

and this is compared to the tabled value of F, with ν_1 and ν_2 degrees of freedom. H_0 is rejected when $F > F(\nu_1 : \nu_2)$.

Two way analysis of variance

Statistical applications

As the name implies, two way analysis of variance deals with two non-random sources of variation, as opposed to the one non-random source in the one way version. The general principle remains the same — to break down total variance into components associated with the two explanatory variables, plus an error term.

Assumptions

The assumptions given for the one way case also apply to the two way version. However a fourth assumption must be added:

4. The effects due to the sources of variation are *additive.* This is a key assumption deserving of some elaboration by way of a hypothetical example. Suppose turnips are being grown, and two sources of variation are considered, namely the treatment of plots with phosphates and with nitrates. The model states that the mean yield of turnips in plots treated with both fertilizers is the grand average yield *plus* the effect due to phosphate treatment, *plus* the effect due to nitrate treatment, *plus* a random error term. More formally, then

$$\bar{X}_{ij} = \bar{\bar{X}} + P_i + N_j + e_{ij} \qquad \qquad \textbf{10.7}$$

When *interactions* are present, then relationships are not additive. Just suppose, for instance, that when nitrates and phosphates are applied simultaneously they cause turnips to shrivel up: if this were to happen, the additivity assumption would be violated by the unexpected inter-action of the chemicals.

Test statistic

Suppose we have data on which we want to apply two way analysis of variance, and there are two categories for Variable I, three categories for Variable II, and four observations in each box, as is illustrated in Table 10.3. In the classic experimental situation Variable I might be treatment of plots with phosphates, II with nitrates, and X records the yield of a crop. An untreated control plot is normally included in such experiments. Notice that the Xs in Table 10.3 have three subscripts, i, j and k: i records the categories (A or B) for Variable I; j records the categories (R, S or T) for Variable II; and k records the number of observations within each box (sub-class).

Using Table 10.3 for reference, one needs the sum of the Xs, and the sum of the Xs squared, for the following:

1. for each sub-class   (i.e. $\sum_k X_{ijk}$ and $\sum_k X_{ijk}^2$);

2. for each row   (i.e. $\sum_j \sum_k X_{ijk}$ and $\sum_j \sum_k X_{ijk}^2$);

3. for each column   (i.e. $\sum_i \sum_k X_{ijk}$ and $\sum_i \sum_k X_{ijk}^2$);

4. for the whole table   (i.e. $\sum_i \sum_j \sum_k X_{ijk}$ and $\sum_i \sum_j \sum_k X_{ijk}^2$).

Some quantities must now be defined. Three lower case parameters are:

ℓ = the number of categories for Variable I ($i = 1, \ldots, \ell$)
m = the number of categories for Variable II ($j = 1, \ldots, m$)
n = the number of observations in each box ($k = 1, \ldots, n$)

As before,

N = the grand total number of observations = $\ell \times m \times n$.

In Table 10.3, $\ell = 3$, $m = 2$, $n = 4$ and $N = 24$.

With this information, seven formulae can now be presented, each one identifying a different type of variation. First, the total sum of the squares (SS_T)

$$SS_T = \sum_i \sum_j \sum_k X_{ijk}^2 - \frac{\left(\sum_i \sum_j \sum_k X_{ijk} \right)^2}{N} \qquad \textbf{10.8}$$

The sum of the squares between the rows, that is, the sum of the squares associated with Variable I, is SS_R where

Table 10.3: Tabular representation of two way analysis of variance

Category		Variable II			Row totals
		R	S	T	
Variable I	A	X_{111} X_{112} X_{113} X_{114} $\sum_k X_{11k}$ $\sum_k X^2_{11k}$	X_{121} X_{122} X_{123} X_{124} $\sum_k X_{12k}$ $\sum_k X^2_{12k}$	X_{131} X_{132} X_{133} X_{134} $\sum_k X_{13k}$ $\sum_k X^2_{13k}$	$\sum_j\sum_k X_{1jk}$ $\sum_j\sum_k X^2_{1jk}$
	B	X_{211} X_{212} X_{213} X_{214} $\sum_k X_{21k}$ $\sum_k X^2_{21k}$	X_{221} X_{222} X_{223} X_{224} $\sum_k X_{22k}$ $\sum_k X^2_{22k}$	X_{231} X_{232} X_{233} X_{234} $\sum_k X_{23k}$ $\sum_k X^2_{23k}$	$\sum_j\sum_k X_{2jk}$ $\sum_j\sum_k X^2_{2jk}$
Column totals		$\sum_i\sum_k X_{i1k}$ $\sum_i\sum_k X^2_{i1k}$	$\sum_i\sum_k X_{i2k}$ $\sum_i\sum_k X^2_{i2k}$	$\sum_i\sum_k X_{i3k}$ $\sum_i\sum_k X^2_{i3k}$	Grand totals $\sum_i\sum_j\sum_k X_{ijk}$ $\sum_i\sum_j\sum_k X^2_{ijk}$

$$SS_R = \frac{\sum\limits_{i}\left(\sum\limits_{j}\sum\limits_{k} X_{ijk}\right)^2}{\ell} - \frac{\left(\sum\limits_{i}\sum\limits_{j}\sum\limits_{k} X_{ijk}\right)^2}{N} \qquad \textbf{10.9}$$

The corresponding sum of the squares for Variable II, representing the sum of the squares between the columns (SS_C) is

$$SS_C = \frac{\sum\limits_{j}\left(\sum\limits_{i}\sum\limits_{k} X_{ijk}\right)^2}{m} - \frac{\left(\sum\limits_{i}\sum\limits_{j}\sum\limits_{k} X_{ijk}\right)^2}{N} \qquad \textbf{10.10}$$

The fourth of these components is the between sub-class sum of the squares, SS_B. This is the sum of the squares due to the rows and columns jointly.

$$SS_B = \frac{\sum\limits_{i}\sum\limits_{j}\left(\sum\limits_{k} X_{ijk}\right)^2}{n} - \frac{\left(\sum\limits_{i}\sum\limits_{j}\sum\limits_{k} X_{ijk}\right)^2}{N} \qquad \textbf{10.11}$$

The interaction sum of the squares, which is the fifth type of variation, is given by

$$SS_I = SS_B - (SS_R + SS_C) \qquad \textbf{10.12}$$

This term is the sum of the squares due to the rows and columns jointly minus the sums of the squares due to the rows and columns individually: it is important in testing the additivity assumption of the model.

As will be demonstrated shortly, the interaction term is tested for significance at the outset by comparing it to the sum of the squares for the within sub-class errors (SS_{ES}) where

$$SS_{ES} = SS_T - SS_B \qquad \textbf{10.13}$$

If the interaction term is not significant then it is conventionally added into the error term. The seventh, and final type of variation, the error sum of the squares, (SS_E) is given by

$$SS_E = SS_T - (SS_R + SS_C) = SS_{ES} + SS_I \qquad \textbf{10.14}$$

Testing is first done for SS_I, in the hope that this term is insignificant. A simple table, as set out in Table 10.4, is recommended for this purpose.

The F value for the interaction test is the ratio of the variance estimates for the interaction effect and the within sub-class errors, thus

$$F = \frac{\hat{s}_I^2}{\hat{s}_{ES}^2} \qquad \textbf{10.15}$$

Table 10.4:   Testing for interaction effects

Source of variation	Sums of squares	Degrees of freedom	Estimate of variance	F
Interaction	SS_I	$(\ell-1)(m-1)=\nu_1$	$SS_I \div \nu_1 = \hat{s}_I^2$	$\hat{s}_I^2 \div \hat{s}_{ES}^2$
Error within sub-class	SS_{ES}	$N - \ell m = \nu_2$	$SS_{ES} \div \nu_2 = \hat{s}_{ES}^2$	

which is compared to $F(\nu_1 : \nu_2)$ and the numbers of degrees of freedom are as defined in Table 10.4.

A complication arises in using this test for interaction in tables which have only one observation in each box (i.e. $n = 1$) since in this special, but not uncommon, case there is no within sub-class error. In this instance, a partitioned chi square test may be applied if the data consist of absolute frequencies, but otherwise a visual inspection of the error term for each sub-class is required. Provided that no obvious interactions are present, then the interaction effect is treated as error, and in the case where $n = 1$, SS_I becomes SS_E.

Having tested for interaction, one proceeds to test the two variables for their effects. Again, a tabulation set out like Table 10.5 is recommended.

Table 10.5:   Testing for the effects of variables one and two in two way analysis of variance.

Source of variation	Sums of squares	Degrees of freedom	Estimate of variance	F
Total	SS_T	$N - 1$		
Variable I	SS_R	$\ell - 1 = \nu_1$ (I)	$SS_R \div \nu_1 \text{ (I)} = \hat{s}_R^2$	$\hat{s}_R^2 \div \hat{s}_E^2$
Variable II	SS_C	$m - 1 = \nu_1$ (II)	$SS_C \div \nu_1 \text{ (II)} = \hat{s}_C^2$	$\hat{s}_C^2 \div \hat{s}_E^2$
Error	SS_E	$N - \ell - m + 1 = \nu_2$	$SS_E \div \nu_2 = \hat{s}_E^2$	

Hence for Variable I, the variance ratio is

$$F = \frac{\hat{s}_R^2}{\hat{s}_E^2} \qquad\qquad \textbf{10.16}$$

which is compared to the F value in Table H, pages 260–61, with $\nu_1 = \ell - 1$ and $\nu_2 = N - \ell - m + 1$. The corresponding variance ratio for Variable II is

$$F = \frac{\hat{s}_C^2}{\hat{s}_E^2}$$ **10.17**

which is compared to F_α with $m - 1$ and $N - \ell - m + 1$ degrees of freedom. As in all these tests, H_0 is rejected when the computed value of F is greater than F_α.

Example

Productivity has for a long time been used as a measure of manufacturing performance, although only recently have economic geographers turned their attention to regional productivity. This example uses aggregated Canadian data for 1970 drawn from a study in progress (Norcliffe and Mitchell, in preparation). The measure to be used is labour productivity, defined as the value added by a manufacturing industry divided by the total number of employees (excluding head office employees) in that industry. Two sources of variation are considered: first, regional variations, and second, the mix of industries. The manufacturing categories (Variable II) that will be used are:

Group R: Wood-based industries
Group S: Textile, clothing and leather
Group T: Minerals, metals and engineering
Group U: Resource based industries including food, petroleum, chemicals and tobacco.

For Variable I the five regions to be used are:

A: Atlantic provinces.
B: Quebec.
C: Ontario.
D: Prairie provinces.
E: British Columbia.

The raw data are presented in Table 10.6, while the sums of the Xs and the sums of the squares for each sub-category, row, column and for the whole table is given in Table 10.7.

PROBLEM DEFINITION

1. H_0: mix and regional effects do not account for significant proportions of productivity variations in Canada.

Table 10.6:   Labour productivity (in $000s per employee) by region and industry group in Canada, 1970

Region	Industry Group R	S	T	U
A	11·468	5·298	9·879	8·960
	6·912	6·879	7·769	18·497
	10·275	3·104	12·272	20·482
B	12·279	7·150	14·278	14·272
	8·665	9·203	11·925	14·797
	12·878	6·673	14·453	14·564
C	12·456	7·335	14·204	15·822
	9·820	10·165	15·356	18·126
	12·461	6·882	15·082	13·674
D	12·190	6·267	12·552	13·356
	8·320	8·937	9·580	18·196
	11·391	6·732	18·522	23·898
E	13·321	7·844	14·232	14·552
	10·023	7·851	11·034	13·800
	12·333	6·816	15·412	17·538

H_1: there are significant variations in productivity amongst the five regions, taking industrial mix into account.

H_2: there are significant variations in productivity amongst the four groups of industries, taking the regional effect into account.

2. $\alpha = ·05$.

TEST SELECTION

1. Productivity is measured at the ratio scale.

2. $\ell = 5$; $m = 4$; $n = 3$; $N = 60$.

3. The additivity assumption will be considered first by testing for interaction. Table 10.8 presents the values corresponding with the quantities in Table 10.4 (the calculation for these values will be presented shortly).

Table 10.7:   Sums of the Xs and X squared for sub-classes, rows, columns and grand totals*

| Region | Industry Group | | | | |
	R	S	T	U	Row totals
A	28·655 (284·87)	15·281 (85·02)	29·920 (308·55)	47·939 (841·93)	121·795 (1520·38)
B	33·822 (391·70)	23·026 (180·35)	40·656 (554·96)	43·633 (634·75)	141·137 (1761·75)
C	34·737 (406·86)	24·382 (204·49)	44·642 (665·03)	47·622 (765·87)	151·383 (2042·24)
D	31·901 (347·57)	21·936 (164·47)	40·654 (592·39)	55·450 (1080·59)	149·941 (2185·02)
E	35·677 (430·01)	22·511 (169·62)	40·678 (561·83)	45·890 (709·78)	144·756 (1871·25)
Column totals	164·792 (1861·01)	107·136 (803·95)	196·550 (2682·76)	240·534 (4032·92)	709·011 (9380·64)

* Sums for values of X squared are enclosed in parentheses. Values are subject to rounding errors.

Table 10.8:   Testing the Canadian productivity data for interaction effects

Source of variation	Sums of squares	Degrees of freedom	Estimate of variance	F
Interaction	46·91	$4 \times 3 = 12$	3·909	0·562
Error within sub-class	278·24	$60 - 20 = 40$	6·956	

Since the F value for interaction is less than unity, it cannot possibly be significant, and a comparison to a tabled F value is unnecessary. No significant interaction effects are present and the additivity assumption holds.

There is not the space to present a detailed examination of the assumptions concerning the error terms. Suffice it to say that although these assumptions may in practice be relaxed, there is no need to do so in this case since the raw data approximately meet the requirements of normality, with mean of zero and equal variance. Given the non-experimental nature of the data, meeting the random sampling assumption involves accepting the premise that a stochastic process is being measured.

SAMPLING DISTRIBUTION

Taking the first research hypothesis, $v_1 = 4$, and $v_2 = 52$: consulting Table H, page 260, $F_{.05} (4,52) = 2.55$. For the second research hypothesis, $v_1 = 3$, and $v_2 = 52$: consulting Table H, $F_{.05} (3,52) = 2.79$.

COMPUTATION

Each of the sums of the squares given by formulae **10.8** to **10.14**, inclusive, will be evaluated. Beginning with the total sum of the squares,

$$SS_T = 9380.64 - \frac{709.011^2}{60} = 9380.64 - 8378.29 = 1002.35$$

The sum of the squares for the rows, which reflects the variation due to the five regions, is

$$SS_R = \frac{121.795^2 + 141.137^2 + 151.383^2 + 149.941^2 + 144.756^2}{5} - 8378.29$$

$$= 8425.58 - 8378.29 = 47.29$$

The sum of the squares for the columns, which is the variation attributable to the four industrial categories, is

$$SS_C = \frac{164.792^2 + 107.136^2 + 196.550^2 + 240.534^2}{4} - 8378.29$$

$$= 9008.20 - 8378.29 = 629.91$$

The between sub-class sum of the squares was used in the test of interaction (above).

$$SS_B = \frac{28.655^2 + 15.281^2 + \ldots + 40.678^2 + 45.890^2}{3} - 8378.29$$

$$= 9102.40 - 8378.29 = 724.11$$

The between sub-class sum of the squares is used to obtain the interaction sum of the squares

$$SS_I = 724 \cdot 11 - (47 \cdot 29 + 629 \cdot 91) = 46 \cdot 91$$

The interaction test involves comparing SS_I to the within sub-class errors, where

$$SS_{ES} = 1002 \cdot 35 - 724 \cdot 11 = 278 \cdot 24$$

The seventh and final sum of the squares, SS_E , is the within sub-class errors plus the interaction sum of the squares.

$$SS_E = 278 \cdot 24 + 46 \cdot 91 = 325 \cdot 15$$

Proceeding directly to testing the two variables for their effects, the data and results are summarized in Table 10.9.

Table 10.9: Results of two way analysis of variance for the Canadian example

Source of variation	Sums of squares	Degrees of freedom	Estimate of variance	F
Total	1002·35	59		
Variable I (regions)	47·29	4	11·82	1·89
Variable II (mix)	629·91	3	209·97	33·60
Error	325·15	52	6·25	

Taking H_1, the F value in this case is given by **10.16**:

$$F = \frac{11 \cdot 82}{6 \cdot 25} = 1 \cdot 89$$

Since $F_{.05}$ (4,52) = 2·55, the calculated F value is not significant and H_1 cannot be accepted: the productivity variations amongst the five regions is not significant.

In the case of H_2, applying **10.17**

$$F = \frac{209 \cdot 97}{6 \cdot 25} = 33 \cdot 60$$

which is very much higher than the critical F value ($F_{.05}$ (3,52) = 2·79): H_2 is accepted and we conclude that there are highly significant productivity variations amongst the four groups of industries.

Geographical applications

Analysis of variance techniques have been applied quite extensively within geography. Probably the most frequent application is to the study of the internal structure of cities, particularly in empirical tests of Hoyt's model. According to this model, socio-economic status varies both between sectors that radiate from the city centre, and between zones within sectors with status rising as distance from the city centre increases. Hence the two sources of variation are sectors and zones. Two short papers by Johnston (1970) and Murdie (1970) deal with some interesting problems that stem from this type of application. In his original study, Murdie (1969) conducted two-way analysis of variance of economic status in sectors and in zones of Toronto in 1951. Murdie found a significant effect for the six sectors he was using, but not for the six zones. Johnston then applied one way analysis of variance to each of six sectors and in three cases found a significant zonal effect. It would seem that two way analysis of variance may sometimes mask effects that can be detected in the one way case, and vice versa. Of equal interest are the problems arising from the significant interaction effect present in Murdie's data. It was recommended earlier that an insignificant interaction effect be included with the error sums of the squares. In re-working Murdie's data, Johnston used the estimate of variance for the interaction effect $(\hat{s}_I^2)$ as the denominator of his F ratios, whereas Murdie (1970) used the within sub-class error variance estimate $(\hat{s}_{ES}^2)$ and obtained much higher F values. There is some ambiguity as to which is correct, but Blalock's (1960, Chapter 16) guidelines suggest that Murdie probably used the better measure.

Conceptually similar to Murdie's work is the study of land values in Topeka by Knos (1962). Since Knos was examining sectoral variations in urban land values, he used one way analysis of variance. Davis (1971), on the other hand, was concerned with the influence of income, age and the white/non-white ratio on variations in housing values in Washington DC. The resulting three-way analysis of variance is complicated by first order interactions between each pair of variables, and second-order interactions amongst all three variables acting together. Other examples of this technique are listed in Greer-Wootten (1972, Section 17).

A difference of variance test for two independent samples

Statistical applications

The purpose of this test is very similar to that of the two sample difference of means tests dealt with in the previous chapter, except that the parameter under consideration is not the mean but the variance. The question posed is whether two sample variances could have been drawn from the same population; notationally, does $\sigma_1^2 = \sigma_2^2$?

This test has obvious statistical uses since there are several tests in which equal variances are assumed, for instance, the point biserial coefficient of correlation (Chapter 7) and the third of the four difference of means tests in Chapter 9. It is sometimes known as the variance ratio test.

Assumptions

1. $\hat{s}_1^2$ and $\hat{s}_2^2$ are independent estimates of σ.
2. The populations from which the samples are drawn are normally distributed.

It is worth mentioning that no assumptions are made either about the sample size, or about the means of the population.

Test statistic

To conduct this test, four values are required, the variances of the two samples ($\hat{s}_1^2$ and $\hat{s}_2^2$) and the number of observations in each sample (N_1 and N_2). With this information one can compute F where

$$F = \frac{\hat{s}_1^2}{\hat{s}_2^2} \qquad \qquad \textbf{10.18}$$

and this is compared to the tabled F value where $v_1 = N_1 - 1$ and $v_2 = N_2 - 1$.

This test is very simple, but one caution should be noted: the F table is set up in such a way that only values of F larger than 1·0 are included. This means that the *larger variance should be placed in the numerator* of equation **10.18**.

Example

The simplicity of this test makes a formal presentation of an example unnecessary. Instead, a quick statistical check will be made of three earlier examples in which equality or non-equality of variance was asserted.

CASE 1

In Chapter 7, data for Snohomish county were used to illustrate the point biserial correlation coefficient. The two samples consisted of towns with and without general stores: using the same notation, the sample variances and sizes were

$$\hat{s}_0^2 = 0.318 \qquad N_0 = 18$$

$$\hat{s}_1^2 = 0.373 \qquad N_1 = 15$$

Since $\hat{s}_1^2$ is the larger value, it becomes the numerator in **10.18** and

$$F = \frac{0.373}{0.318} = 1.173$$

As $F_{.05}(14, 17) = 2.33$, the computed variance ratio is much less than the critical value and the assumption of equal variances is supported.

CASE 2

In Chapter 9, contrasts in industrial growth rates between the American heartland and hinterland were considered in a difference of means test. It was asserted that $\sigma_1^2 = \sigma_2^2$. The sample data was

$$\hat{s}_1^2 = 4090.98 \qquad N_1 = 22$$

$$\hat{s}_2^2 = 4537.80 \qquad N_2 = 27$$

Placing the larger variance in the numerator of **10.18**,

$$F = \frac{4537.80}{4090.98} = 1.109$$

This value is much less than the critical value ($F_{.05}(26, 21) = 2.03$) and again the assumption of equal variances is sustained.

CASE 3

The last of the four tests presented in Chapter 9 was a difference of means test for two samples assuming that $\sigma_1^2 \neq \sigma_2^2$: the data were drawn from interviews in Kitchener—Waterloo. We have

$$\hat{s}_1^2 = 1\cdot932 \qquad\qquad N_1 = 68$$

$$\hat{s}_2^2 = 0\cdot828 \qquad\qquad N_2 = 45$$

$$F = \frac{1\cdot932}{0\cdot828} \quad 2\cdot33$$

Consulting Table H with $v_1 = 67$ and $v_2 = 44$, by interpolation $F_{.05} = 1\cdot60$ and $F_{.01} = 1\cdot94$, hence the difference between the two sample variances is highly significant, as was assumed.

Geographical applications

This test is rarely used in the direct examination of geographical problems. However, geographers frequently need to compare variances as a prerequisite to other tests (there are three instances in this text), hence the test ought to be in fairly common use as an intermediate step in geographical analysis.

11 Pearson's product-moment coefficient of correlation

Statistical applications

This chapter is concerned with the correlation coefficient known as Pearson's r. It is probably fair to say that this is the single most important statistical measure in use, its importance being attributed to, first, it being the most powerful and generally used measure of correlation, and second, it being the basic ingredient in the majority of multivariate techniques.

Since Pearson's r is a 'standardized measure of linear covariance', a formal statistical definition of covariance is required. If there are two interval scale variables, X and Y, then the covariance of X and Y is

$$cov\ (XY) = \frac{\Sigma\ (X_i - \bar{X})\ (Y_i - \bar{Y})}{N} = \frac{\Sigma x_i y_i}{N} \qquad 11.1$$

If one considers **11.1** carefully, it will be evident that covariance differs from variance in one important respect: the latter cannot be a negative quantity whereas the former can. Thus, for instance, as the temperature rises, so the sale of ice cream increases and the two variables co-vary positively; but as the temperature rises so the sale of overcoats declines and these two variables have negative covariance. The term 'standardized' implies that the coefficient has a standard range from +1 to −1 regardless of the magnitude of the variables involved.

Like the measures of correlation discussed in earlier chapters, a value of $r = +1$ indicates perfect positive correlation, while $r = -1$ indicates perfect negative association. A value of zero indicates that the two variables are statistically independent. r is a sample parameter, and is commonly used to make inferences about ρ (rho) which is the corresponding population parameter.

In many textbooks correlation and regression are discussed in the same chapter — and with good reason since from a statistical viewpoint they are closely related. Nevertheless from the user's viewpoint there are important differences between the two methods so that in this text they will be treated in separate chapters. To be specific, whereas in regression one assumes a *causal* relationship between X and Y in correlation one simply assumes *association*. A significant association between two variables can be attributed to at least four mechanisms.

1. X and Y are both dependent variables, influenced by variable A. Symbolically:

$$A$$
$$\swarrow \searrow$$
$$X \qquad Y$$

2. X and Y are linked through one or more intermediary variables, A_i, to form a causative chain, hence, when $i = 2$,

$$X \rightarrow A_1 \rightarrow A_2 \rightarrow Y$$

3. X causes Y to change, but there is a feedback loop from Y to X so that two-way interrelationships obtain.

$$X \rightleftharpoons Y$$

4. A one-way causal relationship of the type assumed in regression analysis is operating:

$$X \rightarrow Y$$

Besides, as in all inferential tests, there is a probability corresponding to α of making a Type I error, i.e. rejecting the null hypothesis when it is true. It follows that one should not make causative inferences on the basis of simple correlation analysis, since two variables may appear to be associated for a number of different reasons.

In this text, we are only concerned with the pairwise correlation of X and Y. When more than two variables are being correlated, multiple and partial correlation techniques may be applied. These methods are treated in most intermediate texts, and a clear exposition with geographical examples may be found in Yeates (1974).

There is, of course, a big difference between explanation in terms of real world processes and statistical explanation. In parametric statistics, statistical explanation is frequently reckoned in terms of variance explained. Although correlation coefficients should be treated with care in 'explaining' the real world, they provide a very simple measure of statistical explanation. The square of the correlation coefficient, which is known as the *coefficient of determination* (r^2) is a direct measure of the proportion of the variance in a bivariate distribution explained by a linear correlation coefficient. For instance, if $r^2 = \cdot77$, then 77% of the total variance is 'accounted' for by a linear correlation. Although the coefficient of determination is obtained from the product moment coefficient, it is used mainly with regression methods in assessing the explanatory power (in a statistical sense) of a given regression model.

Assumptions

1. Values should be recorded at the interval or ratio scale.

2. There are two variables, X and Y, which have a joint distribution which is bivariate normal. Two terms in the preceding sentence deserve some elaboration. A 'joint distribution' is one in which two variables are plotted simultaneously against each other, as in Figure 11.1. A bivariate normal distribution is one in which both the *marginal* and the *conditional* distributions of X and Y are normal. A marginal distribution

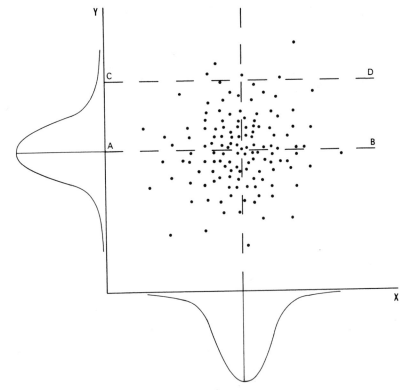

Figure 11.1: A bivariate normal distribution.

is the frequency distribution obtained by summing the distribution of a variable on to its respective axis: thus in Figure 11.1 the normal curves drawn against the X and Y axes are theoretical marginal distributions for these two variables. The conditional distribution is normal when, for any value of X, Y is normally distributed, and for any value of Y, X is normally distributed.

Figure 11.1 shows a bivariate normal distribution with zero correlation. The observations form a roughly circular pattern with an increasing density of points towards the centre of the distribution such that if one were to generalize the distribution to a continuous surface it would be dome shaped. Notice that the generalized marginal distributions are normal. Not only are both marginal distributions normal, but any horizontal or vertical line drawn through the swarm of points would, in theory, yield a conditional distribution that is normal. Hence the transect $A-B$ corresponding with $\bar{Y}$ should have a normal

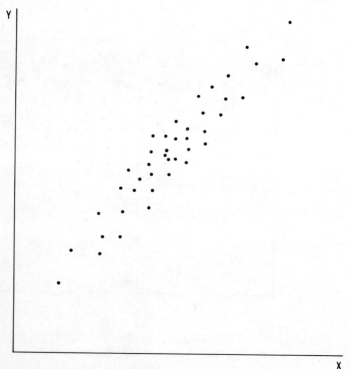

Figure 11.2: A bivariate normal distribution with strong positive correlation.

distribution for X, as should the transect $C-D$ and any other horizontal or vertical transect.

Diagrams of bivariate normal distributions provide a useful means of illustrating the meaning of linear correlation. Figure 11.2 shows a bivariate distribution with strong positive correlation. Notice that in the theoretical distribution, either a horizontal or a vertical transect would give rise to a normal curve. However when a relationship is present the swarm of points is not circular but stretched into an ellipsoid shape, and the stronger the correlation, the greater is the amount of stretching and the larger is the ratio of the long axis of the ellipse to the short axis.

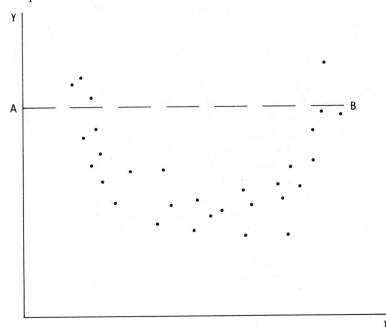

Figure 11.3: A bivariate distribution with zero linear correlation but a strong curvilinear relationship.

Figure 11.3 shows a distribution with a strong curvilinear correlation, although the linear correlation is close to zero. The purpose of this diagram is to show that if there is a curvilinear relationship then the distribution is not bivariate normal: for instance, the transect $A-B$ in Figure 11.3 yields a bimodal conditional distribution.

It should be added that the assumption of a bivariate normal distribution may be relaxed somewhat. As in the case of analysis of variance, experimental evidence suggests that the normality assumption need not be interpreted too rigidly, especially for large samples, although for small samples it is more important.

3. The question of sample size in correlation is a little more complicated. It will be considered first, for the hypothesis that $\rho = 0$, and second, for the hypothesis that $\rho >$ or < 0, or for comparing two sample coefficients.

(i) Testing whether $\rho = 0$.

Applying Gaito's statement (quoted in Chapter 1) to simple linear correlation, we deduce that, provided the parent distribution is bivariate normal, inferences may be drawn from samples of any size: the greater unreliability of smaller samples is taken into account in tables of critical values.

The caveat concerning the bivariate normality of the parent distribution is important. Consider the distributions in Figures 11.4 and 11.5, both showing bivariate distributions for samples consisting of 10 points. In Figure 11.4, there are 9 points which are almost totally uncorrelated, but point a in the upper right part of the graph induces a fairly strong overall correlation. With such a small sample it is not possible to assess whether the parent distribution is normal: the sample does not look very normal, but with only 10 observations such a distribution could be quite easily drawn from a bivariate normal population. The reverse is true in Figure 11.5: point b has the effect of almost reducing to zero the fairly strong correlation present in the other 9 observations. If these sample distributions were found to be drawn from positively skewed populations and a logarithmic transformation were applied, then the sample correlation coefficients for the untransformed and the transformed data would differ quite radically, especially for Figure 11.4. Put briefly, the problem reduces to this: one needs a sample with something in the order of 30 observations before one can assess whether the assumption of a bivariate normal distribution is met. However this does not apply if there are grounds for arguing that the sample is drawn from a bivariate normal population: such grounds may be provided either by parallel empirical examples, or by theoretical arguments. Hence if one is confident that the population is bivariate normal, then the 'null null hypothesis' may be tested with quite small samples.

Gaito's statement relates to Type I errors (rejecting a null hypothesis

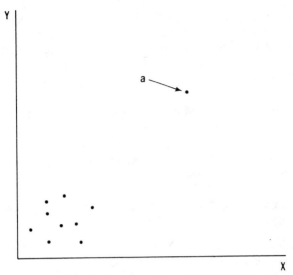

Figure 11.4: Correlation in small samples: a fairly high correlation attributable to one observation.

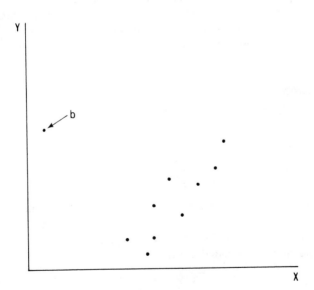

Figure 11.5: Correlation in small samples: an otherwise fairly high correlation made much lower by a single observation.

that is true). Given a bivariate normal population, the critical values give α protection against making a Type I error with samples of any size. The weakness of small samples is that the probability of making a Type II error (failing to validate a true research hypothesis) increases sharply with small samples. Hence the tests become progressively less powerful with small samples.

It is sometimes suggested that when ambiguous distributions such as those in Figures 11.4 and 11.5 are encountered, further sampling should take place. In experimental situations this may be sound advice, but geographers often work with non-experimental problems which makes repeat sampling impossible.

(ii) Testing whether $\rho >$ or < 0, or comparing two sample correlation coefficients.

When ρ is close to zero, the sampling distribution of r is approximately normally distributed. However the sampling distribution of r becomes skewed as ρ diverges from zero, and is highly skewed when ρ is close to +1 or −1, so that tests that use the standard error of r are not reliable in these circumstances. What is needed is some way of normalizing the sampling distribution of r. The required transformation producing an *approximately* normal form was developed by R. A. Fisher and is known as Fisher's Z. The approximation is good for samples as large as 50, indeed the transformation produces a tolerably normal distribution for samples as small as 15. However, for very small samples the transformed sampling distribution is less acceptably normal.

Test statistic

Since the Pearsonian correlation coefficient is a standardized measure of covariance, it is appropriate that the formula for r first be presented in these terms:

$$r = \frac{cov\,(XY)}{\sqrt{var\,(X)\,var\,(Y)}} = \frac{cov\,(XY)}{s_X\,s_Y} \qquad \textbf{11.2}$$

where $cov\,(XY)$ is the covariance of XY as defined in **11.1**, and s_X and s_Y are the standard deviations of X and Y respectively. Since each of the quantities in **11.2** is divided by N, the denominators cancel each other out and we can re-write **11.2** more fully as

$$r = \frac{\Sigma(X_i - \bar{X})(Y_i - \bar{Y})}{\sqrt{\left[\Sigma(X_i - \bar{X})^2\right]\left[\Sigma(Y_i - \bar{Y})^2\right]}} = \frac{\Sigma xy}{\sqrt{(\Sigma x^2)(\Sigma y^2)}} \qquad \textbf{11.3}$$

The version of r given in **11.3** illustrates fairly clearly why Pearson's coefficients are also known as the 'product-moment' coefficient. Moments were defined in Chapter 3: recall that the rth moment about the mean (π_r) is given by

$$\pi_r = \frac{\Sigma (X_i - \bar{X})^r}{N} \qquad \textbf{3.15}$$

Applying this, and cancelling out the Ns, then the numerator of **11.3** is the product of the first moments of X and Y, while the denominator is the square root of the product of the second moments of X and Y.

Although formulae **11.2** and **11.3** are very literal equations the most commonly used machine formula is presented in equation **11.4**.

$$r = \frac{N \Sigma X_i Y_i - (\Sigma X_i)(\Sigma Y_i)}{\sqrt{\left[N \Sigma X_i^2 - (\Sigma X_i)^2\right]\left[N \Sigma Y_i^2 - (\Sigma Y_i)^2\right]}} \qquad \textbf{11.4}$$

Notice that this equation involves only six quantities, five of which are summations, as follows:

1. N, the total number of observations;
2. the sum of the Xs;
3. the sum of the Ys;
4. the sum of the Xs squared;
5. the sum of the Ys squared;
6. the sum of the product of X and Y.

A table set up as follows makes it very easy to obtain these six quantities:

Observation	X	Y	X^2	Y^2	XY
1	–	–	–	–	–
2	–	–	–	–	–
.	.	.	.	.	.
.	.	.	.	.	.
.	.	.	.	.	.
N	–	–	–	–	–
TOTALS	ΣX	ΣY	ΣX^2	ΣY^2	ΣXY

Hence in most circumstances **11.4** provides the simplest method of calculating r.

Various significance tests can be applied to correlation coefficients: the three most commonly used tests will be considered here.

TESTING THE NULL HYPOTHESIS THAT $\rho = 0$

In this test, a sample correlation coefficient is compared to a theoretical population in which under the null hypothesis ρ is assumed to be zero. Most commonly one hopes that the null hypothesis will be rejected. The test statistic is identical to that used with Spearman's r_s and the point biserial coefficient:

$$t = r\sqrt{\frac{N-2}{1-r^2}} \qquad\qquad 11.5$$

and t is compared to the tabled value of Student's t with $N-2$ degrees of freedom. Alternatively, tables have been published which permit the significance of r to be assessed directly: these are reproduced in a number of textbooks (eg. Walker and Lev, 1969, Table VI).

TESTING THE NULL HYPOTHESIS THAT $\rho >$ or < 0.

In this test one might, for example, be testing the 'null' hypothesis that $\rho = \cdot 6$ against an H_1 that $\rho > \cdot 6$. First convert both ρ and r using Fisher's Z transformation where

$$Z = 1 \cdot 1503 \log_{10}\left(\frac{1+r}{1-r}\right) \qquad\qquad 11.6$$

(in the latter case substitute ρ for r). Next, calculate the standard error of Z from

$$s_Z = \frac{1}{\sqrt{N-3}} \qquad\qquad 11.7$$

One then obtains the standard normal deviate, z (not to be confused with Fisher's Z) from

$$z = \frac{Z_r - Z_\rho}{s_Z} \qquad\qquad 11.8$$

and this is compared to the tabled value of z in Table F, page 258; H_0 is rejected when $z > z_\alpha$.

COMPARING TWO SAMPLE CORRELATION COEFFICIENTS

In this case the test is to determine whether two correlation coefficients could have been drawn from the same population. Notationally, does $r_A = r_B$? For this test, both correlation coefficients must be transformed using Fisher's Z transformation, **11.6**, to obtain two Z values, Z_A and Z_B. Also needed is the standard error of the Z values in each sample, given by **11.7**. One then computes the standard deviation of the sampling distribution of the difference in Z values, from

$$s_{Z_A - Z_B} = \sqrt{s_{Z_A}^2 + s_{Z_B}^2} \qquad \qquad \textbf{11.9}$$

which is, in turn, used to obtain a z value, where

$$z = \frac{Z_A - Z_B}{s_{Z_A - Z_B}} \qquad \qquad \textbf{11.10}$$

z is then compared to the critical value, z_α, in Table F.

Example

This example will take up one small part of the 'development' problem, namely the relationship between diet and income in various countries. The data is drawn from a paper by Pedro Belli (1971) who used United Nations publications as his source. The two variables to be correlated are as follows:

1. average total proteins (in grams) consumed per caput per day in 1958;
2. average income (in $) per caput in 1968.

The reason for the ten-year time lag between the two variables will become apparent when the two variables are regressed in Chapter 13.

Belli's original data would allow thirty-seven pairs of observations. However an examination of the distribution of the countries involved showed irregular representation. Missing were countries from Eastern Europe and Africa south of the Sahara, while Western Europe and Latin America were well represented. The over-representation of Latin America and the under-representation of Africa were seen as roughly cancelling each other out, leaving the most developed nations over-represented. Hence six of the twelve wealthiest countries were randomly dropped, leaving a sample of thirty-one countries with a fairly balanced mix of developed and developing countries. Clearly, the sample is not random in the formal sense, but at least the most obvious bias in favour of the developed countries was corrected. The data is given in Table 11.1

PROBLEM DEFINITION

1. H_0: there is no correlation between average protein consumption in 1958 and average income in 1968 ($\rho = 0$).

H_1: Average protein consumption in 1958 and average income in 1968 are positively correlated with each other ($\rho > 0$).
The test is one-tailed.

$\alpha = \cdot01$

TEST SELECTION

1. Both variables are measured at the ratio scale.
2. $N = 31$.
3. Of the three assumptions, the first is clearly met with the data measured at the ratio scale. The second assumption, that of a bivariate normal distribution, required some investigation. First, histograms of both variables were plotted (see Figure 11.6 and 11.7). The histogram of the marginal distribution for the protein consumption variable is not normal, but nor is it skewed, indeed the distribution is fairly rectangular. No transformations were therefore applied to the protein consumption data which is defined as variable X in its raw form. The income variable, in contrast, has a strong positive skew. Computing the standardized

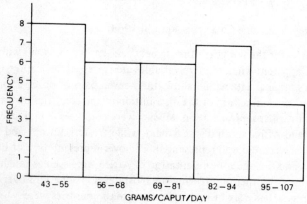

Figure 11.6: Histogram of raw data on protein consumption (variable X).

Table 11.1: Per caput daily supply of proteins (1958) and per caput income (1968)

		Total proteins consumed (1958) (Grams/cap./day) X	Income 1968 ($ per cap.)	Log income Y	X²	Y²	XY
1	Argentina	98	519	2·715	9604	7·37	266·1
2	Austria	87	1104	3·043	7569	9·26	264·7
3	Brazil	61	252	2·401	3721	5·77	146·5
4	Canada	95	2087	3·320	9025	11·02	315·4
5	Chile	77	465	2·667	5929	7·11	205·4
6	Columbia	48	262	2·418	2304	5·85	116·1
7	Denmark	92	1947	3·289	8464	10·82	302·6
8	Greece	95	651	2·814	9025	7·92	267·3
9	India	51	73	1·863	2601	3·47	95·0
10	Ireland	96	840	2·924	9216	8·55	280·7
11	Israel	83	1158	3·064	6889	9·39	254·3
12	Italy	77	1020	3·009	5929	9·05	231·7
13	Japan	67	928	2·968	4489	8·81	198·8
14	Libya	53	926	2·967	2809	8·80	157·2
15	Mexico	68	478	2·679	4624	7·18	182·2
16	Netherlands	77	1465	3·166	5929	10·02	243·8
17	New Zealand	105	1714	3·234	11025	10·46	339·6
18	Norway	84	1673	3·223	7056	10·39	270·8
19	Pakistan	46	108	2·033	2116	4·14	93·5
20	Paraguay	68	192	2·283	4624	5·21	155·3
21	Peru	49	241	2·382	2401	5·67	116·7
22	Philippines	47	233	2·367	2209	5·60	111·3
23	Portugal	71	423	2·626	5041	6·90	186·5
24	Spain	71	707	2·849	5041	8·12	202·3
25	Sri Lanka	45	133	2·124	2025	4·51	95·6
26	Syria	46	203	2·307	2116	5·32	106·1
27	Taiwan	57	221	2·344	3249	5·50	133·6
28	Turkey	73	299	2·476	5329	6·13	180·7
29	Venezuela	61	761	2·881	3721	8·30	175·8
30	United Kingdom	86	1560	3·193	7396	10·20	274·6
31	United States	92	3303	3·519	8464	12·38	323·7
	Totals	2226		85·151	169940	239·22	6293·9

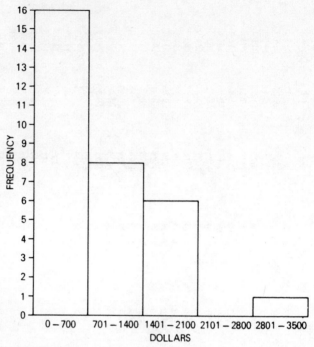

Figure 11.7: Histogram of raw data on incomes.

coefficients of skewness and kurtosis, β_1 and β_2 respectively, (as
defined in equations **3.17** and **3.19**), we obtain:

	β_1	β_2
Protein consumption (Variable X)	·00054	1·77
Income	2·00	5·02
Income − log transformed (Variable Y)	·0407	2·14

Consulting Table G, page 259, it is clear that a β_1 of 2·00 for a sample
of thirty-one observations is significantly skewed. A log transformation
was therefore applied to the raw income data, and the result, as the
histogram in Figure 11.8 demonstrates, was virtually to eliminate the
skewness (β_1 falls to ·0407). For both variable X, and for the transformed income data, β_2 is less than 3 indicating platykurtic marginal
distributions. The log transformation of income was adopted, and
from this point all reference to variable Y implies the logarithmic
transformation of this variable. The bivariate distribution was then
plotted in Figure 11.9: although the resulting plot is not perfectly

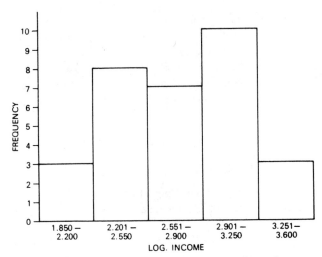

Figure 11.8: Histogram of income data − log transformed (variable Y).

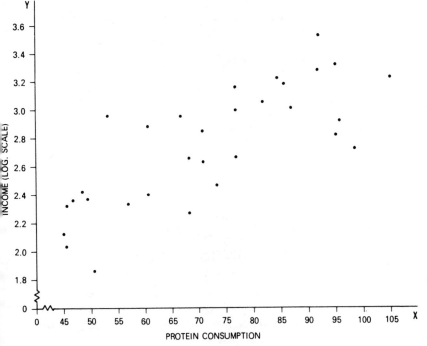

Figure 11.9: Bivariate distribution of X and Y.

bivariate normal, the marginal distributions are not seriously non-normal. There is not sufficient data to give a clear picture of the conditional distributions, but there is no visual evidence of non-normality. The third assumption relates to sample size: for the purpose of testing the null hypothesis that $\rho = 0$, a sample of thirty-one observations is adequate. Pearson's r will therefore be computed and then tested using a t test.

SAMPLING DISTRIBUTION

$N = 31$, so $\nu = N - 2 = 29$. Consulting Table E, page 257, with $\alpha = \cdot01$ and 29 degrees of freedom in a one-tailed test, we find $t_{\cdot01} = 2\cdot46$.

COMPUTATION

The bases of the computations are the column totals in Table 11.1. Using **11.4** we obtain

$$r = \frac{(31)(6293\cdot9) - (2226)(85\cdot151)}{\sqrt{\left[(31)(169940) - 2226^2\right]\left[(31)(239\cdot22) - 85\cdot151^2\right]}}$$

$$= \frac{195110\cdot9 - 189546\cdot13}{\sqrt{[5268140 - 4955076][7415\cdot82 - 7250\cdot69]}}$$

$$= \frac{5564\cdot77}{\sqrt{[313064][165\cdot13]}}$$

$$= \frac{5564\cdot77}{7190\cdot01} = \cdot774$$

Testing r for significance according to **11.5**

$$t = \cdot774\sqrt{\frac{31 - 2}{1 - \cdot599}} = \cdot774\sqrt{\frac{29}{\cdot401}}$$

$$= \cdot744\sqrt{72\cdot319} = 6\cdot582$$

which far exceeds the critical value of $2\cdot46$. Hence we conclude that there is a highly significant positive correlation between daily protein consumption in 1958 and income in 1968.

The above conclusion is based on a sample of thirty-one observations.

Belli conducted a similar test using a sample of thirty-seven observations and obtained a correlation of ·721. The two r values can be compared by using Fisher's Z transformation as defined in **11.6**. Using the subscript A for the above results and B for Belli's results, then

$$Z_A = 1·1503 \log_{10} \left(\frac{1 + ·774}{1 - ·774}\right)$$

$$= 1·1503 \log_{10} \left(\frac{1·774}{·226}\right) = 1·1503 \log_{10}(7·8496)$$

$$= (1·1503)(0·8948) = 1·0302$$

$$Z_B = 1·1503 \log_{10} \left(\frac{1 + ·721}{1 - ·721}\right) = ·9097$$

The variances are obtained by squaring the values in **11.7**:

$$s^2_{Z_A} = \frac{1}{31 - 3} = ·0357$$

$$s^2_{Z_B} = \frac{1}{37 - 3} = ·0294$$

The pooled estimate of the standard deviation of the sampling distribution of the differences in Z values, as defined in **11.9**, becomes

$$s_{Z_A - Z_B} = \sqrt{·0357 + ·0294} = ·2552$$

Finally, applying **11.10**,

$$z = \frac{1·0302 - ·9097}{·2552} = ·4772$$

Consulting Table F, page 258, which gives probabilities associated with different z values under the normal curve, the computed value has a probability of ·63 of occurring, therefore we conclude that the difference between the two sample correlations is insignificant.

Suppose, instead, that we had tested the null hypothesis that $\rho = ·5$ versus an H_1 that $\rho > ·5$. In this case **11.6**, **11.7** and **11.8** are evaluated. $Z_r = 1·030$ and $Z_\rho = ·549$. Using **11.7** the standard error is ·189, and evaluating **11.8**

$$z = \frac{1·030 - ·549}{·189} = \frac{·481}{·189} = 2·55$$

In this case we conclude that r is significantly greater than ·5 at the ·01 level of significance.

Geographical applications

Correlation techniques are now so widely used in geography that it is not uncommon to find that up to half of the articles in a geographical journal employ correlation analysis in one form or another. Some of the applications employ correlation analysis for descriptive rather than inferential purposes, in the sense that the aim is quantitatively to examine the relationships between variables without formally testing any hypotheses. However, the majority of more recent papers draw inferences from correlation analysis, and with good reason since it is good practice to assess the significance of a coefficient.

Geographers commonly wish to measure correlation in areally based data, which introduces two important problems. First, correlations based on modifiable units such as towns, census tracts and counties are such that if the boundaries of these areal units are changed the correlations are also likely to change. Related to this is the weighting issue illustrated by the classic problem of correlating wheat yield with rainfall. Should counties be weighted according to their area so that big counties influence the size of the correlation coefficient more than small counties do? The problem of modifiable units will be considered more fully in the final chapter. The second problem concerns the use of raw data versus ratios (such as percentages). For instance, McCarty *et al.* (1956) used raw data in correlating the locational patterns of various manufacturing industries: nearly all the resulting coefficients were high positive values because of the uneven population of his areal units, with manufacturing concentrated in larger towns. Kuh and Mayer (1955) have examined the use of ratios as an alternative: there are problems with this, including the danger of spurious correlations, but for cross-sectional employment data this is not likely to be very serious. By expressing the employment in each industry in a town as a proportion of the total employment in that town, Richter (1969) has virtually eliminated the problem of inflated correlations encountered by McCarty *et al*.

One particular geographical application of correlation that has the prime virtue of simplicity is a grouping technique known as *elementary linkage analysis.* This technique, as developed by McQuitty (1957) uses correlation coefficients descriptively to give a quantitative measure of similarity between two variables. However, it can be given an inferential basis by allowing only those coefficients which are significant (at some appropriate level) to be used in the grouping procedure. When the variables to be grouped are geographical areas, then linkage analysis becomes an algorithm for areal grouping or regionalizing. The technique will be illustrated by a simple example drawn from a study by Norcliffe (1968). Various unemployment characteristics (such as levels of unemployment, male and female unemployment, seasonal variations, and age differences in the unemployed) for eleven employment areas in southwest England in 1967–8 were correlated with each other. The correlations are given in Table 11.2.

The grouping procedure (incorporating the significance criterion) involves the following five steps:

1. Working at an appropriate level of significance identify the critical value of *r*. In the example, a minimum value of ·62 was required for a correlation to be significant.

2. Scan the columns of the correlation matrix, in each case identifying the largest positive *r* in each column: circle the largest *r* if it is greater than the critical value of *r*. The largest *r* in each column of Table 11.2 was found to be significant and was therefore circled.

3. Identify the largest positive *r* in the entire matrix. This *r* will link two variables which form a *reflexive pair* and consequently become the core of Group I. In the example, the largest *r* is ·85 which links St Austell (F) and Wadebridge (I) (see Figure 11.10).

4. The procedure now switches to a row-wise search. Scan the row corresponding to both members of the reflexive pair: if a circled value other than the circle identifying the reflexive pair is encountered, then the variable in the column concerned is joined to the group by the row variable. If a member is added to a group in this way, the row for the new group member is then searched, and if further circled values are encountered then these variables are added to the group and their respective rows are searched. This continues until all the linkage routes of a group lead to rows in which no circled values are encountered.

In the example, the rows of the reflexive pair, F and I, are searched. The only circled value in row F is I, the other member of the reflexive pair: however in row I a circled value is encountered for K and Camelford joins the group through I. Switching to row K, a circled value is found for J and Liskeard joins the group through K. There are no circles in Row J, so Group I is complete (see Figure 11.10).

5. One then identifies the highest *r* in the matrix involving variables not yet assigned to a group, and this becomes a new reflexive pair. Step 4 is then repeated, and so on for successive reflexive pairs until every variable with a significant *r* is assigned to a group.

Table 11.2: Correlations between eleven employment areas in southwest England based on unemployment characteristics

		A	B	C	D	E	F	G	H	I	J	K
Newquay	A	—	(73)	−09	·51	·06	·06	·25	·54	·35	·45	·56
Penzance	B	(73)	—	·32	(75)	·29	−01	·50	(84)	·30	·68	·46
Redruth	C	−09	·32	—	·24	(80)	·68	·39	·23	·71	·67	·47
Falmouth	D	·51	·75	·24	—	·20	·13	·18	·52	·38	·63	·36
Helston	E	·06	·29	(80)	·20	—	·54	·30	·26	·76	·77	·77
St Austell	F	·06	−01	·68	·13	·54	—	·23	−04	(85)	·54	·58
Truro	G	·25	·50	·39	·18	·30	·23	—	·79	·16	·64	·31
Bodmin	H	·54	(84)	·23	·52	·26	−04	(79)	—	·09	·65	·44
Wadebridge	I	·35	·30	·71	·38	·76	(85)	·16	·09	—	·75	(81)
Liskeard	J	·45	·68	·67	·63	·77	·54	·64	·65	·75	—	·80
Camelford	K	·56	·46	·47	·36	·77	·58	·31	·44	·81	(80)	—

SOURCE: Norcliffe (1968).

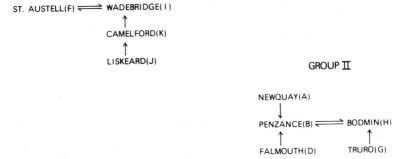

GROUP I

ST. AUSTELL(F) ⇌ WADEBRIDGE(I)

CAMELFORD(K)

LISKEARD(J)

GROUP II

NEWQUAY(A)

PENZANCE(B) ⇌ BODMIN(H)

FALMOUTH(D)   TRURO(G)

GROUP III

REDRUTH(C) ⇌ HELSTON

Figure 11.10: Groups of employment areas in southwest England defined by elementary linkage analysis.

In the example, the eleven employment areas were collapsed into the three groups shown in Figure 11.10.

Needless to say, elementary linkage analysis has its limitations: in particular, a single link ties a member to a group so that it is possible, through chaining, for quite unlike members to belong to a group. Despite these limitations, it is a very simple technique to operate: moreover quite frequently it gives rise to groups whose membership is very similar to groups defined by more sophisticated grouping procedures.

Two other geographical versions of correlation techniques, namely ecological correlations and spatial autocorrelation, will be considered in the final chapter of this book.

12 Some tests of spatial order: testing for directional preference and for contiguity

In his paper seeking to identify some fundamental spatial concepts, Nystuen (1963) suggested *distance, direction,* and *connection* as fundamental or 'primitive' terms. Papageorgiou (1969) has subsequently revised this list to include *point* and *time.* In Chapter 6 it was established that one of these primitives, namely *point* patterns, may be analysed statistically. This chapter is concerned with two more of these primitive terms, namely *direction* or *orientation,* and *contiguity* (a form of connection). As with most statistical methods, one can rapidly move into complex problems of inference, but there are available some straightforward tests that may be used to identify certain basic aspects of directionality and contiguity. Two tests are described here: the first is a test for the existence of a preferred orientation in a sample. The second is a test of contiguity for mosaic maps with two colours.

Identifying an orientational preference: the Rayleigh test

Statistical applications

The present test is named after the eminent mathematician Lord Rayleigh (J. W. Strutt) who, as early as 1880, was publishing papers on the pitch and phase of vibrations. Rayleigh's work dealt with random vibrations in one, two and three dimensions: however in this chapter we are concerned only with two dimensional problems, hence the aim is to identify an orientational preference on a plane.

The Rayleigh test, and much of the work dealing with orientation on a plane, is based on the circular normal distribution (occasionally known as von Mises's distribution). Circular distributions are a special kind of two-dimensional distribution where the frequency or probability is spread out on the circumference of a circle. Thus, if one cuts out a sketch of a normal distribution and joins the two tails together to make a cylinder, one creates a circular distribution known as the wrapped normal distribution. The circular normal is another circular distribution, and is used in most statistical analyses of directional data.

There are three key parameters of the circular normal distribution. The first is the angle where the distribution peaks — this is known as the mean angle: this can be any angle between 0° and 360°. The second is the relevant measure of dispersion known as the mean angular deviation. The third parameter indicates the shape of the curve. Just as with other distributions, the circular normal distribution may be peaked (leptokurtic) or flat (platykurtic). At one extreme it is completely flat and degenerates into a uniform distribution which resembles a tin can with its top and bottom removed. At the other extreme it becomes very peaked with the whole distribution concentrated along one small part of the circumference of a circle.

The Rayleigh test is designed to identify a preferred orientation. There is available another simple test to determine whether or not a set of angles is uniformly distributed, namely a version of the chi square test. Given a set of angles, one groups these into an appropriate number of arcs of equal size to obtain the observed distribution. The expected distribution, which is uniform in shape, is obtained by dividing N (the number of angles) by k (the number of arcs) and performing the usual chi square test with $k - 1$ degrees of freedom. This version of the chi square test which is described in Doornkamp and King (1971, p. 348) tells you whether you have a significant deviation from a uniform distribution, but it does *not* identify the preferred orientation and hence is not considered to be as useful as the Rayleigh test.

Assumptions

The sample should be drawn from a population with a circular normal distribution. An explicit extension of this assumption is that the distribution should be unimodal. Figure 12.1 gives two instances where it would

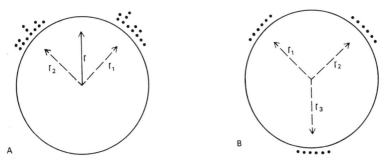

Figure 12.1: Two instances where the Rayleigh test is inappropriate.

be misleading to apply the Rayleigh test. In A, vector r identifies a preferred orientation, whereas there is in practice a bimodal distribution
with two preferred directions corresponding with r_1 and r_2. In B no
preferred direction is found, whereas there are three clusters of observations at r_1, r_2 and r_3.

Test statistic

Before describing the test statistic, a brief resumé of some trigonometry
is needed.

In trigonometry the origin from which angles are measured is east.
Furthermore, the size of an angle increases as one moves anti-clockwise
(contrary to geographic systems). As a result, north, to the geographer
is $0°$ and west is $270°$: but to the mathematician, these orientations are

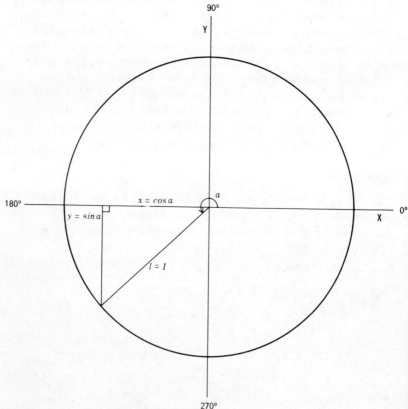

Figure 12.2: Rectangular and polar co-ordinates.

respectively 90° and 180°. Angles shown in Figure 12.2 are those used in trigonometry.

Second, it is useful to distinguish between rectangular co-ordinates and polar co-ordinates. Rectangular or Cartesian co-ordinates are the type used on graph paper where x records the position of the X axis (the abscissa) and y the position on the Y axis (the ordinate). These are illustrated in Figure 12.2. Given an origin, then x and y determine an angle. It is also possible to express an angle in polar co-ordinates, with a as the angle and ℓ the length of the vector from the origin to the point x, y. It is often convenient, in analysing directional data, to treat each angle as a unit vector (i.e. $\ell = 1$). This simply means that each angle is given equal weight.

Before performing the Rayleigh test, two parameters of the *mean vector* are needed, the angle, a, and the length, ℓ. The angle of the mean vector can be obtained by algebraic or geometric methods. The geometric method is simple but tedious: one joins each unit vector end-to-end. The end point of the final vector is then connected back to the origin, this connecting line being the *resultant* and its angle is the same as the angle of the mean vector: this method is illustrated in Figure 12.3 in the example below. The algebraic method of obtaining a is preferred because the computations can be used to obtain several statistical measures. Unfortunately, circles have peculiar properties such that the obvious algebraic solution, the 'mean', as defined in Chapter 3, is not a reliable measure of the mean angle. For example, given three angles, 359°, 1° and 3°, then the arithmetic mean, 121°, is clearly not the mean angle. For this reason, use is made of trigonometry to obtain the angle and the length of the mean vector.

First one computes V, the sum of the cosines of all the angles,

$$V = \Sigma \cos a_i \qquad \textbf{12.1}$$

and W, the corresponding sum of the sines

$$W = \Sigma \sin a_i \qquad \textbf{12.2}$$

The resultant, R, is given by

$$R = \sqrt{V^2 + W^2} \qquad \textbf{12.3}$$

To obtain the parameters of the mean vector, one simply divides those of the resultant vector by N. Hence the rectangular co-ordinates of the mean vector are

$$x = \frac{V}{N} \qquad\qquad 12.4$$

$$y = \frac{W}{N} \qquad\qquad 12.5$$

The length of the mean vector is

$$\ell = \frac{R}{N} \qquad\qquad 12.6$$

The angle of the mean and resultant vectors are identical, hence

$$\cos a = \frac{x}{\ell} = \frac{V}{R}$$

or

$$a = \cos^{-1} \frac{x}{\ell} = \cos^{-1} \frac{V}{R} \qquad\qquad 12.7$$

Given this information, the computation of the Rayleigh test is extremely simple. All one has to do is compute the test statistic

$$H = \frac{R^2}{N} = N\ell^2 \qquad\qquad 12.8$$

and compare it to the table of critical values, H_α, in Table I in the Appendix, page 262. If $H > H_\alpha$ we reject the null hypothesis and conclude that a preferred orientation is present.

One other useful statistic is the mean angular deviation, s_A, which is the circular equivalent of the standard deviation. Once we have computed the length of the mean vector, ℓ, then the mean angular deviation is easily obtained by

$$s_A = 114 \cdot 592 \, (1 - \ell) \quad \text{(degrees)} \qquad\qquad 12.9$$

Some researchers have used the standard deviation rather than the mean angular deviation as a measure of dispersion for circular distributions. Batchelet (1965) indicates that when the range is less than $60°$ and the distribution is fairly symmetrical, then these two measures are very similar. However these two measures diverge when the range exceeds $60°$ or if some skewness is present hence the mean angular deviation should be used in preference to the standard deviation.

Example

One of the problems examined by geographers interested in orientation
is the azimuth of cirques. This example is taken from Evans (1969) and
relates to the orientation of thirty-seven cirques in Jotunheim, Norway.
It is generally held that solar heating plays an important role in the
formation of cirques, hence the expectation that cirque orientations
will be concentrated in those directions which receive least insolation.

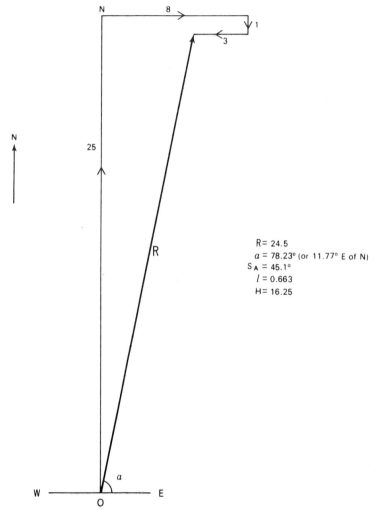

R = 24.5
$a = 78.23°$ (or 11.77° E of N)
$S_A = 45.1°$
$l = 0.663$
H = 16.25

Figure 12.3: Cirque orientation in the Jotunheim (after Evans, 1969).

In the case of Jotunheim, twenty-five of the cirques face to the north, eight to the east, one to the south and three to the west. (See Figure 12.3.)

PROBLEM DEFINITION

1. H_0: the cirques do not display an orientational preference.
   H_1: a preferred orientation is present in the cirques of the Jotunheim.
2. $\alpha = \cdot 05$

TEST SELECTION

1. We assume the data are measured at the interval scale, although angles appear to be generalized to the nearest cardinal direction.
2. $N = 37$.
3. The distribution of the angles is unimodal, and it would seem reasonable to assume that they are drawn from a circular normal population. This being so, the requirements of the Rayleigh test are met.

SAMPLING DISTRIBUTION

Some critical values of the test statistic, H, are given in Table I in the Appendix, page 262. For $N = 37$ and $\alpha = \cdot 05$, the critical value is approximately 2·97, hence if the computed value exceeds this amount, and the mean angle is in the hypothesized direction, we can reject the null hypothesis.

COMPUTATION

The simplest way of organizing the computations is shown in Table 12.1.

Table 12.1: Computation of various circular statistics for cirques in Jotunheim (see text)

Angle	Sine	Cosine	Frequency (f)	$f \times$ sine	$f \times$ cosine
0°	0	+1·0	8	0	+8·0
90°	+1·0	0	25	+25·0	0
180°	0	−1·0	3	0	−3·0
270°	−1·0	0	1	−1·0	0
Total			37	+24·0	+5·0

From this table we have

$$V = \Sigma \cos a_i = 5.0$$

$$W = \Sigma \sin a_i = 24.0$$

Applying **12.3** $$R = \sqrt{25 + 576} = \sqrt{601} = 24.51$$

$$x = \frac{V}{N} = \frac{5.0}{37} = 0.135$$

$$y = \frac{W}{N} = \frac{25.0}{37} = 0.676$$

$$\ell = \frac{R}{N} = \frac{24.51}{37} = 0.663$$

$$\cos a = \frac{x}{\ell} = 0.2038$$

and $a = 78.23°$ (or, using geographical co-ordinates, $11.77°$ degrees east of north)

Applying **12.8** $$H = N\ell^2 = 37 \times .439 = 16.25$$

$$s_A = 114.592 \, (1 - \ell)° = 45.05°$$

H far exceeds the critical value, and the mean angle, $12°$ east of north, is associated with low levels of solar heating at high latitudes in the northern hemisphere. We therefore reject the null hypothesis and conclude that cirques in the Jotunheim have a preferred orientation.

It is worth noting that the mean angular deviation for the Jotunheim data is $45.1°$, whereas the standard deviation is $55.7°$: clearly these two measures of dispersion differ substantially in this case, and the standard deviation is considered to be misleadingly large.

Geographical applications

The majority of empirical studies of orientation have dealt with non-geographical phenomena. The greatest number of applications have been in biology and in geology (Batchelet (1965) refers to many of these). The few geographical applications have dealt with two main problems. First, the study of asymmetry in landform features in general and cirques in particular. The example presented earlier taken from a study by Evans (1969) is one such application, while Unwin (1973) has explored a similar problem in Snowdonia, Wales. As Evans (1972) makes explicit, the purpose of these studies is to make inferences about processes using formal evidence. This returns to the point that geography is an observational rather than an experimental science. Glaciologists cannot stage a repeat of the ice ages, but they can piece together residual landform evidence in a jigsaw-like manner in order to gain an understanding of what processes were at work and how they modified the landscape.

The second major application of directional statistics in geography is to the study of migration. Both Wolpert (1967) and Adams (1969) have examined this problem, using descriptive statistics, but Greer-Wootten and Gilmour (1972) have adopted a more rigorous approach by applying the inferential statistics discussed above to migration patterns within Montreal.

Another application which employs a modification of these statistics is a study of the preferred orientation of Iroquoian Indian longhouses in Ontario by Norcliffe and Heidenreich (1974). In this case, angles are unique only over a 180 degree range: for instance a longhouse that is orientated southeast is also oriented northwest. This problem is overcome by a device attributable to Krumbein (1939) known as the 'doubling of angles'. Angles are measured in the range 0–180 degrees and are then doubled to convert them to a 360 degree range. Hence 65 degrees becomes 130 degrees, and 153 degrees becomes 306 degrees and so on. Once this is done, the statistical methods outlined above are applicable. The longhouse study revealed a significant preference for a northwest–southeast orientation. By comparing this preferred orientation with various climatic data, it was theorized that this orientation is designed to maximize the thermal efficiency of longhouses, particularly on cold and windy days in winter.

Dacey's test for contiguity

Statistical applications

Imagine any sort of mosaic map, such as the counties of England and Wales. Here we will consider the simplest case with counties coloured either black or white. These colours may represent any form of binary variable, where conventionally black represents the 'positive' side and white the 'negative' side. By way of example, we might shade black

1. counties in which a specific activity is present;

2. counties for which the score on a continuous variable is above the mean or the median;

3. counties in which the residuals from a regression are positive (these are discussed in Chapter 13);

4. counties which are functionally linked to some internal or external node.

The question that is posed in a contiguity test is: 'Are the black or the white counties arranged in a definite pattern that is non-random or are the black and white counties randomly distributed?' As the black or white counties become increasingly clumped together, so a point is reached where the level of contiguity is deemed significant. At the other extreme, and this may be more difficult to grasp, the black and white counties may be so arranged that they tend to alternate, in which case we have the opposite of clustering. A random distribution lies somewhere between these extremes, with some random clustering and some alternation. Figure 12.4 illustrates these tendencies and may help clarify the distinction between clustering, randomness and alternation.

Since the data is measured at the nominal scale — black or white — the test is not parametric. However the test statistic is compared to the normal distribution so that is resembles such procedures as the normal approximation to the binomial distribution. It therefore lies part way between the strictly nonparametric test of contiguity presented by Dacey (1968) and some fully parametric tests discussed in Cliff and Ord (1969).

Using B to represent black and W to represent white, and using the term 'join' for two counties with a common border, then the contiguity test involves comparing the number of BB joins, WW joins and BW joins on a map to those expected under the null hypothesis. It is not difficult to see that if all the black counties are clustered in one part of a map and all the white counties in another, then you have more BB and WW joins than expected and fewer BW joins. The contiguity tests described below were developed by Dacey (1968). There are in fact three closely related tests — for BB joins, for WW joins and for BW joins: all three are usually applied.

Referring back to the hypothetical mosaic map shown in Figure 12.4, the number of BB, WW and BW joins for the clustered, random and alternating pattern are given in Table 12.2. In the clustered arrangement there are only 5 BW joins, and most are BB or WW; in contrast, with the alternating arrangement twenty out of twenty-six joins are BW. The random pattern falls between these two extremes.

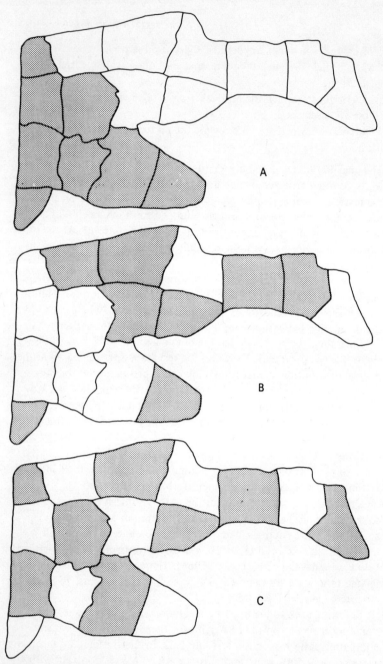

Figure 12.4: Hypothetical mosaic map showing clustering (A),
randomness (B) and alternation (C) in a two-colour system.

Table 12.2: Number of joins of various types for Figure 12.4.

		Joins			
		BB	WW	BW	Total
A:	Clustered	10	11	5	26
B:	Random	6	8	12	26
C:	Alternating	3	3	20	26

The contiguity test described below is probably the simplest test of *spatial autocorrelation*. The meaning of the term 'autocorrelation' is self explanatory; just as the author of an autobiography is writing about himself, so an autocorrelated variable is correlated with itself. The term has had greatest use in time series analysis, particularly by econometricians examining cyclical fluctuations, and the alternative term *serial correlation* is also used. Spatial autocorrelation is simply a variable that is correlated with itself through space. Hence the clustered pattern in Figure 12.4A is an example of positive spatial autocorrelation; for any given county shaded black, there is a good chance that a neighbouring county will also be shaded black, and likewise for white counties. In contrast, the alternating pattern in Figure 12.4C is an example of negative spatial autocorrelation because the majority of neighbouring counties are of the opposite colour.

Having stated that the contiguity test is probably the simplest test of spatial autocorrelation, it may be useful to point out three routes along which further elaborations may be found:

1. By increasing the number of colours above two; these are called *k*-colour maps.

2. By measuring special types of contiguity on a regular square grid or lattice: for instance the 'Queen's case' and the 'Rook's Case' where these terms reflect contiguity patterns just as in a game of chess. These are discussed in Cliff (1967).

3. By applying the true parametric tests of contiguity alluded to in the introduction to this section.

Assumptions

The key assumption is whether the three test statistics approach a normal form, so that the normal distribution provides a good approximation for hypothesis testing purposes. Dacey discusses this problem and draws the following tentative conclusions:

1. If the study region is elongated (such as Chile) the number of counties, N, should be large.

2. In the case of a study region that is not highly elongated, the test is applicable if one of the following three conditions obtain:

either   (a)   the number of counties is large

or   (b)   the number of counties is moderately small and p is large (where p is the probability of a cell being black)

or   (c)   the number of counties is moderately large and p not small.

To some extent these assumptions beg the question because the definitions involve such undefined terms as 'large' and 'moderately small'. The following arbitrary and tentative definitions are added by Dacey:

p is small when $p < \cdot 2$;

p is large when $p > \cdot 7$.

Corresponding definitions for counties are not given by Dacey, but in the context of his discussion and examples, a number as large as 100 would appear to be 'large', a value around 20 would appear to be 'moderately small' and 40 would appear to be 'moderately large'.

Test statistic

The first definition is L, representing the total number of joins on a map. If there are N counties, and L_i is the number of links for county i, then the grand total, L, is given by

$$L = \frac{\Sigma L_i}{2} \qquad \qquad \textbf{12.10}$$

The sum is divided by two because each join is counted twice.

Using N_B and N_W to represent the number of counties that are black and white respectively, then the probability of a county being black, p, is

$$p = \frac{N_B}{N} \qquad \qquad \textbf{12.11}$$

and the probability of a county being white, q, is

$$q = \frac{N_W}{N} = 1 - p \qquad \qquad \textbf{12.12}$$

In an irregular lattice, the expected mean number of joins are

for BB joins

$$\mu(BB) = p^2 L \qquad\qquad \textbf{12.13}$$

for WW joins

$$\mu(WW) = q^2 L \qquad\qquad \textbf{12.14}$$

for BW joins

$$\mu(BW) = 2pqL \qquad\qquad \textbf{12.15}$$

To obtain the variances, one other constant needs defining, namely K, where

$$K = \Sigma\, L_i(L_i - 1) \qquad\qquad \textbf{12.16}$$

The expected variances are

for BB joins

$$\sigma^2(BB) = p^2 L + p^3 K - p^4(L + K) \qquad\qquad \textbf{12.17}$$

for WW joins

$$\sigma^2(WW) = q^2 L + q^3 K - q^4(L + K) \qquad\qquad \textbf{12.18}$$

for BW joins

$$\sigma^2(BW) = 2pqL + pqK - 4p^2 q^2(L + K) \qquad\qquad \textbf{12.19}$$

With this information it is easy to calculate the three test parameters:

$$z(BB) = \frac{J(BB) - \mu(BB)}{\sigma(BB)} \qquad\qquad \textbf{12.20}$$

$$z(WW) = \frac{J(WW) - \mu(WW)}{\sigma(WW)} \qquad\qquad \textbf{12.21}$$

$$z(BW) = \frac{J(BW) - \mu(BB)}{\sigma(BW)} \qquad\qquad \textbf{12.22}$$

where $J(BB)$, $J(WW)$ and $J(BW)$ represents the empirically observed number of joins of each type. Tests for randomness are evaluated by treating these quantities as standard normal deviates. Thus if we were conducting a two tailed test at the ·05 level of significance and $z(\text{BB}) = +2\cdot13$ we would consult Table F, page 258, and see that the probability associated with this outcome is less than ·05, hence H_0 is rejected. Normally one is not testing just one of the quantities, but all three.

Example

The example that is developed here is identical to one used by Dacey (1968). It is drawn from Taaffe, Morrill and Gould's (1963) study of transport expansion in underdeveloped countries, and tests whether the residuals from a regression are autocorrelated in space. Taaffe *et al.* fitted an equation to Ghanaian data that predicted road mileage as a function of the population density of a district. Some districts have more miles of roads than predicted and are positive residuals (shaded black) while other have fewer miles and are negative residuals (shaded white in Figure 12.5)

It is worth pointing out that the results presented here differ very slightly from those of Dacey because of the difficulty of reconstructing the original map. The boundaries of two of the districts (Kumasi 1–2 and Kumasi 5) were not drawn on any of the source maps and so had to be approximated; also some of the joins were ambiguous. For the purposes of exposition, we have here taken the liberty of relocating some boundaries slightly so that the all joins are unambiguous.

PROBLEM DEFINITION

   1. H_0: the residuals are randomly distributed.

   H_1: the residuals are autocorrelated in space so that significant contiguity is present.

   2. $\alpha = \cdot05$

TEST SELECTION

   1. A binary variable is used.

   2. $N = 40$
   $N_B = 20$
   $N_W = 20$

   3. The study region is fairly compact, there are forty regions hence N is moderately large, and half the districts are black so p is not small. Hence the requirements of the contiguity test are met.

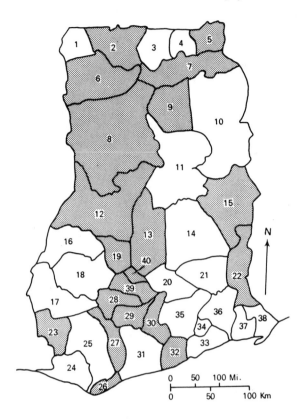

Figure 12.5: Ghana: positive (black) and negative (white) residuals from a regression predicting road mileage as a function of population density.
SOURCE: Taaffe, Morrill and Gould (1963).

SAMPLING DISTRIBUTION

The three statistics $z(BB)$, $z(WW)$ and $z(BW)$ should approach the normal form in this case. We will reject H_0 if a z value is greater than $\pm 1 \cdot 96$.

COMPUTATION

The simplest way of doing the calculations for the contiguity test is to set up a table recording details of all the joins. This is done in Table 12.3.

Table 12.3:   Details of joins for the contiguity test using Ghanaian data

District	Residuals + or −	BB	WW	BW	Total	District	Residuals + or −	BB	WW	BW	Total
1	−	0	0	2	2	21	−	0	4	1	5
2	+	2	0	2	4	22	+	1	0	5	6
3	−	0	1	2	3	23	+	0	0	3	3
4	−	0	1	2	3	24	−	0	1	1	2
5	+	1	0	1	2	25	−	0	2	3	5
6	+	3	0	1	4	26	+	1	0	2	3
7	+	5	0	3	8	27	+	3	0	3	6
8	+	4	0	1	5	28	+	3	0	2	5
9	+	2	0	2	4	29	+	4	0	1	5
10	−	0	1	3	4	30	+	3	0	3	6
11	−	0	2	5	7	31	−	0	0	5	5
12	+	3	0	2	5	32	+	1	0	3	4
13	+	3	0	3	6	33	−	0	3	1	4
14	−	0	3	3	6	34	−	0	3	0	3
15	+	1	0	3	4	35	−	0	5	2	7
16	−	0	2	2	4	36	−	0	5	1	6
17	−	0	3	3	6	37	−	0	2	1	3
18	−	0	2	3	5	38	−	0	1	1	2
19	+	4	0	2	6	39	+	5	0	2	7
20	−	0	3	4	7	40	+	3	0	1	4
						Total		52	44	90	186

First we need

$$p = \frac{N_B}{N} = \frac{20}{40} = \cdot 5$$

and

$$q = 1 - p = 1 - \cdot 5 = \cdot 5$$

We also need the two constants, L and K.

$$L = \frac{\Sigma L_i}{2} = \frac{186}{2} = 93$$

$$K = \Sigma \left(L_i \left(L_i - 1 \right) \right) = 776$$

The expected number of joins of each type are

$$\mu(\text{BB}) = p^2 L = \cdot 25 \times 93 = 23 \cdot 25$$

$$\mu(\text{WW}) = q^2 L = \cdot 25 \times 93 = 23 \cdot 25$$

$$\mu(\text{BW}) = 2pqL = \cdot 5 \times 93 = 46 \cdot 5$$

and their respective variances are

$$\sigma^2(\text{BB}) = p^2 L + p^3 K - p^4(L + K)$$

$$= 23 \cdot 25 + 97 - 54 \cdot 31 = 65 \cdot 94$$

$$\sigma^2(\text{WW}) = q^2 L + q^3 K - q^4(L + K) = 65 \cdot 94$$

$$\sigma^2(\text{BW}) = 2pqL + pqK - 4p^2q^2(L + K)$$

$$= 46 \cdot 5 + 194 - 217 \cdot 25 = 23 \cdot 25$$

With this information we now obtain the three test statistics:

$$z(\text{BB}) = \frac{J(\text{BB}) - \mu(\text{BB})}{\sigma(\text{BB})} = \frac{26 - 23 \cdot 25}{8 \cdot 12} = \cdot 34$$

$$z(\text{WW}) = \frac{J(\text{WW}) - \mu(\text{WW})}{\sigma(\text{WW})} = \frac{22 - 23 \cdot 25}{8 \cdot 12} = - \cdot 15$$

$$z(\text{BW}) = \frac{J(\text{BW}) - \mu(\text{BW})}{\sigma(\text{BW})} = \frac{45 - 46 \cdot 5}{4 \cdot 82} = - \cdot 31$$

All three computed values are quite close to zero: we therefore must accept the null hypothesis and conclude that no significant contiguity is present in the pattern of residuals.

Geographical applications

Insofar as the contiguity test is explicitly a method for spatial analysis, it was logical to allude to certain uses of the test in discussing its statistical applications. In identifying certain criteria for shading a county black or white, four such uses were mentioned: these will now be elaborated upon.

The first use is to test whether there is a tendency for an activity or phenomenon to agglomerate in space. For instance in the point biserial coefficient of correlation there are areas where a variable is either present or absent: the same data could be used with the contiguity test and the result will indicate the existence or absence of an agglomerative tendency.

The example presented above has demonstrated the use of the technique in testing for contiguity in the residuals from a regression. The residuals are assumed to independent random variables, a condition which is violated if contiguity is present.

In situations where a continuous variable is examined for spatial autocorrelation it is best to use parametric tests such as those developed by Cliff and Ord. However computation of these statistics for large samples is exceedingly laborious without a computer, whereas the contiguity test described above is easy to calculate. By colouring all counties with scores above the mean or median black and all others white, it is easy to reduce a continuous variable to a binary level. This obviously involves throwing away information for the sake of computational simplicity, but in some situations this is quite justifiable. It is also possible that the amount of information lost is not great, but tests would be required to confirm this.

The fourth use of the contiguity test is in examining a map for functional linkages, and in particular for the influence of proximity on diffusion patterns. Diffusion implies a functional link between an 'adopter' and a node or source which has already adopted. The diffusion of many types of phenomena, especially epidemics and the like, is assumed to be subject to a neighbourhood effect; adoption is more likely to occur near to the source than further away. As a result, adoption patterns should tend to be contiguous in space.

13   Simple linear regression

Statistical application

Regression differs from topics discussed in previous chapters in that it is primarily used as a modelling technique. This is not to say that regression has no inferential basis: on the contrary regression may be used to test hypotheses, and this requires that point and interval estimates be made. However the purpose of most regression models is to go beyond the demonstration of general association between two variables by predicting specific values for one variable in terms of another.

Recall that in correlation the aim was to measure the strength of the linear association between two variables. In regression the relationship is specified such that X is designated an *independent* variable that causes Y, the *dependent* variable, to vary. Symbolically, $Y \leftarrow X$. Very rarely does this mean that the value of Y is entirely determined by

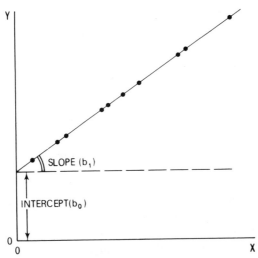

Figure 13.1: A perfect linear relationship.

X, at least when geographical problems are being considered, although perfect linear relationships of the type shown in Figure 13.1 are sometimes encountered in the physical sciences. More commonly Y is only partly explained by X, the result being a relationship more like that shown in Figure 13.2 with a scattering of points around the regression line.

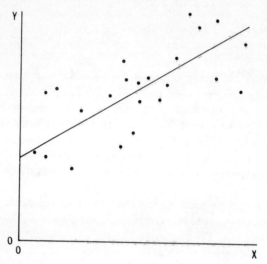

Figure 13.2: A regression line with imperfect predictive power.

The rarity of perfect linear relationships in geography is attributable to two causes:

1. The phenomena which geographers study are usually multivariate in character. By this it is meant that any given variable is influenced by several other variables so that the single independent variable, X, accounts for only a portion of the variation in Y. In such a situation, multiple regression is commonly applied and the single variable X is replaced by a vector of variables, X_i, so that

$$Y \leftarrow X_1, X_2, X_3, \ldots, X_k$$

Multiple regression is not discussed in this text, but it is easy to see that it is a logical extension of simple regression.

2. From a geographical point of view, many of the observable features of the landscape are subject to random or stochastic influences. In other words, although some of the variation in a phen-

omenon may be logically attributed to a set of explanatory variables, there remains a component of the variance which is inherently unpredictable and which is attributed to chance happenings, be they floods, the whims of land speculators, unusually cold weather, or unexpected deaths in a family. The role of stochastic processes is particularly stressed in the work of Curry, and many of his ideas are synthesized in his important essay, 'Chance and landscape' (1967). Conditioned as we are to believe in an ordered landscape, Curry's ideas may seem heretical, yet the fact remains that we are still extraordinarily poor at forecasting geographical aspects of events — in most cases because of chance occurrences. In regression, random disturbances are readily incorporated as part of the error term.

Assumptions

The assumptions of the linear regression model have been examined in an important paper by Poole and O'Farrell (1971). The discussion that follows is based largely on their paper.

The simple linear regression model that is being considered states that:

$$Y_i = b_0 + b_1 X_i + e_i \qquad \textbf{13.1}$$

Where Y is the dependent variable;

   X is the independent variable;

   b_0 is the intercept coefficient, the value of Y at the point where $X = 0$;

   b_1 is the slope coefficient (as illustrated in Figure 13.1);

   e is an error term.

The two coefficients, b_0 and b_1, are estimated using least squares. The regression model may be used for either descriptive of inferential purposes: since this book focuses on inferential statistics, it is the latter use that will be considered.

A distinction should be drawn between the 'fixed X' model and the 'random X' model. In the former case, X is fixed at some predetermined level and the resulting value of Y is recorded; in the latter case, in contrast, X is sampled randomly from a universe of Xs. Suppose one could fix the rainfall falling in different places at X inches a year, and one then recorded wheat yields (Y): if rainfall could be so controlled, then one would have a fixed X model. It should be clear from this illustration that only in exceptional cases do geographers use the fixed

X model, for like most non-experimental sciences, one has very little control over the processes that are being monitored.

The seven assumptions that follow are required by the random X version of the simple linear regression model when it is being used for inferential purposes.

1. The data are measured at the interval or ratio scale.

2. The Xs are measured without error. It is less important that Y be measured without error, although if gross errors and/or systematic errors are present, then the results will be suspect. If error is introduced in the measurement of Y, then the stochastic disturbance term, e, is defined to include this measurement error.

3. The relationship between X and Y is linear in the form the variables are regressed.

4. The value of e (the residuals from the regression) are pairwise uncorrelated. Recent work by Martin (1974) indicates that if spatial autocorrelation is present in e then there is loss of efficiency in estimating the regression coefficients.

The next three assumptions relate to the conditional distribution of the error term, e. Recall that in discussing the assumptions of simple correlation in Chapter 11, a conditional distribution was defined. In this case, the conditional distribution being considered is the frequency distribution of e corresponding with any given value of X.

5. Each conditional distribution of e has a mean of zero.

6. The error term is homoscedastic: that is, each conditional distribution of e has constant variance.

7. Each conditional distribution of e (and therefore of Y) is normally distributed. This is required for tests based on the t distribution and, as in other t tests, as the sample size becomes large, so this assumption may be progressively relaxed.

The above list of assumptions may have a somewhat chilling effect on those anxious to use the linear regression model. Poole and O'Farrell certainly intend this. They point out that the assumption most commonly mentioned by geographers, indeed in many cases the only assumption referred to, is the normality assumption: yet this assumption does not even apply if one is using regression for descriptive purposes by making point estimates of the coefficients b_0 and b_1, and even when the normality assumption does apply, it can often be relaxed because a sample is fairly large. Fortunately, the assumptions listed above tend, in practice, not to be 'independent', in the sense that if one is violated then several others are usually violated. For instance, if the conditional distribution of Y is highly skewed, then there is a good chance that the relationship between X and Y is not linear, and if a linear regression were fitted, the *es* may be serially dependent, and their conditional distributions heteroscedastic with non-zero means. In short, if assumption 7 is not met, then it is quite likely that assumptions 3, 4, 5 and 6 will also not be met. The converse is also likely to be true, so that in normalizing their data geographers have often, and perhaps unwittingly, met several other assumptions.

Fitting a simple regression line

Given a bivariate distribution plotting X against Y, there are several methods of fitting a regression line through the swarm of points. For instance, some very simple yet quite accurate graphical methods are available. The usual approach, however, is to fit a least squares regression line by evaluating the relevant equations.

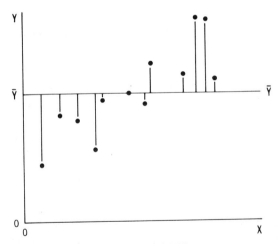

Figure 13.3: Deviations from the mean of Y.

Figures 13.3 and 13.4 are designed to give an intuitive understanding of what 'least squares' implies. The same distribution consisting of twelve points is plotted in both diagrams. In Figure 13.3, the vertical lines show the distance from each point to the arithmetic mean of the Ys. *The aim of least squares is to minimize the square of these deviations,* hence the least squares regression line is plotted in Figure

13.4. Note that since it is the square of the deviations (i.e. the variance) that is minimized, the regression line is pulled towards those points with the largest deviations from the mean.

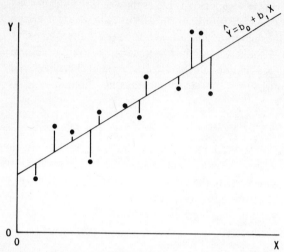

Figure 13.4: Deviations from the least squares regression line of Y *on* X.

In Chapter 3, the estimation of parameters was briefly discussed. One of the conventions used then was to put a hat on certain estimates: in this chapter $\hat{Y}$ denotes the value of Y predicted by a regression equation. A regression equation can therefore be written in two ways: in **13.1** we had

$$Y_i = b_0 + b_1 X_i + e_i$$

or $$\hat{Y} = b_0 + b_1 X \qquad\qquad \textbf{13.2}$$

Equation **13.2** represents what is 'explained' by the regression model, while e is the part that is unexplained. The vertical lines in Figure 13.4 are the es corresponding with each observation.

The unexplained and explained components are illustrated in Figure 13.5. Assume that a negatively sloping regression is fitted through a swarm of points; three observations, P, Q and R will be considered. The solid vertical lines show the deviations of P, Q and R from $\bar{Y}$, while the dotted lines show the residuals (e_i). Interest focuses on the dashed line, which corresponds with the 'explained' component. For Q the deviation from $\bar{Y}$ is reduced by the regression line, but for P the regression line

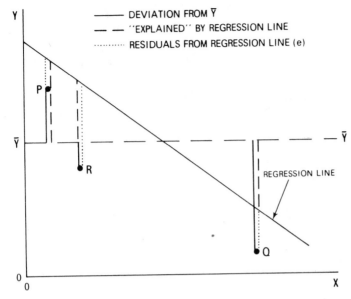

Figure 13.5: Explained and unexplained components in regression.

over-explains so that the dashed line is longer than the original deviation. The situation shown in R may also sometimes arise: the residual from the regression is greater than the original deviation from $\overline{Y}$! This loss of explanation for an individual observation is offset by the overall explanatory power of a regression line.

Having explained graphically the nature of a least squares regression line, a computational solution is required. What is needed is an estimate of the two sample coefficients, b_0 and b_1. The recommended formulae are:

$$b_0 = \frac{(\Sigma\ Y_i)\ (\Sigma\ X_i^2) - (\Sigma\ X_i)\ (\Sigma\ X_i\ Y_i)}{N\ \Sigma\ X_i^2 - (\Sigma\ X_i)^2} \qquad \textbf{13.3}$$

and

$$b_1 = \frac{N\ \Sigma\ X_i\ Y_i - (\Sigma\ X_i)\ (\Sigma\ Y_i)}{N\ \Sigma\ X_i^2 - (\Sigma\ X_i)^2} \qquad \textbf{13.4}$$

It is useful to note the computational advantages of **13.3** and **13.4**. Referring back to the calculations for Pearson's r, the column totals obtained there may also be used here: moreover, the denominators of **13.3** and **13.4** are the same as the first element in the denominator of r (**11.4**), while the numerator of **13.4** is identical to the numerator for r. Hence if r has been evaluated, little additional work is required to calculate the two regression coefficients.

Figure 13.5 was used to describe regression in terms of explained and unexplained variation. These variances can be defined formally: if $\hat{Y}$ is the value of Y predicted by the regression equation corresponding to each value of X, then the explained variance, $\hat{s}_E^2$ is

$$\hat{s}_E^2 = \frac{\Sigma\,(\hat{Y}_i - \bar{Y})^2}{N-2} \qquad \textbf{13.5}$$

and the unexplained variance, $\hat{s}_{Y.X}^2$, is simply the variance of the residuals about the regression line (i.e. the variance of Y given the effect of X):

$$\hat{s}_{Y.X}^2 = \frac{\Sigma\,(Y_i - \hat{Y}_i)^2}{N-2} = \frac{\Sigma\,e_i^2}{N-2} \qquad \textbf{13.6}$$

Note the additive property of these variances: $\hat{s}_Y^2 = \hat{s}_E^2 + \hat{s}_{Y.X}^2$.

The test for significance of the overall explanatory power of a regression equation is, in effect, a test of the slope coefficient b_1, which in turn can be expressed in standardized terms as a correlation coefficient. In order to make clear the statistical connection between correlation and regression, Pearson's r is expressed in **13.7** in terms of explained and total variation:

$$r = \pm\sqrt{\frac{\Sigma\,(\hat{Y}_i - \bar{Y})^2}{\Sigma\,(Y_i - \bar{Y})^2}} = \pm\sqrt{\frac{\text{explained variation}}{\text{total variation}}} \qquad \textbf{13.7}$$

Sampling theory for regression

The normal procedure in simple linear regression is to use sample data to make point estimates of the regression coefficients, b_0 and b_1. Like all such point estimates, b_0 and b_1, are estimates of the corresponding population parameters, β_0 and β_1 (these beta coefficients should not be confused with the standardized measures of skewness and kurtosis discussed in Chapter 3). Since b_0 and b_1 are predicted from sample data, they are subject to sampling variation so that confidence limits may be placed around each estimate. It can be demonstrated that if the seven assumptions listed earlier are satisfied then least squares estimates of b_0 and b_1 are unbiased and efficient estimates of β_0 and β_1.

In order to calculate the standard errors of the regression coefficients and of the $\hat{Y}_i$'s, the standard deviation about the regression line, $\hat{s}_{Y.X}$, is required. This is simply the square root of the quantity defined in **13.6**. The standard error for the intercept coefficient is then obtained from

$$\hat{s}_{b_0} = \hat{s}_{Y.X} \sqrt{\frac{1}{N} + \frac{\bar{X}^2}{\Sigma (X_i - \bar{X})^2}} \qquad \textbf{13.8}$$

The matching standard error for the slope coefficient is

$$\hat{s}_{b_1} = \frac{\hat{s}_{Y.X}}{\sqrt{\Sigma (X_i - \bar{X})^2}} \qquad \textbf{13.9}$$

Significance tests of b_0 and b_1 make use of these two standard errors. Comparing b_0 to some arbitrary value, B_0, and b_1 to B_1, then we have

$$t = \frac{b_0 - B_0}{\hat{s}_{b_0}} \qquad \textbf{13.10}$$

and

$$t = \frac{b_1 - B_0}{\hat{s}_{b_1}} \qquad \textbf{13.11}$$

and in **13.10**, t is distributed as Student's t with $N - 2$ degrees of freedom. **13.11** is also distributed as t with $\upsilon = N - 2$ provided that the true population parameter, $\beta_1 = 0$. The latter test is therefore generally used to test the null hypothesis that $B_1 = 0$, against the research hypothesis that $B_1 \neq 0$ (two-tailed), or $B_1 > 0$ and $B_1 < 0$ (one-tailed). In short, **13.11** is a test of whether X and Y are linearly associated: if the slope of the regression line is not significantly non-zero, then the independent variable, X, has no explanatory power in the form it is regressed.

The third standard error to be defined is the standard error of the regression line corresponding to any value of X.

$$\hat{s}\,\hat{Y} = \hat{s}_{Y.X}\sqrt{\frac{1}{N} + \frac{(X - \bar{X})^2}{\Sigma\,(X_i - \bar{X})^2}}$$ **13.12**

The standard error of estimate of $\hat{Y}$ is used to place confidence limits (i.e. a prediction interval) about the regression line. Unfortunately, as Haworth and Vincent (1974) lament, there are occasions when geographers have used the standard deviation about the regression ($s_{Y.X}$) to represent prediction limits about a regression line when the standard error of the estimates of $\hat{Y}$, $\hat{s}\hat{y}$, is the correct standard error to use. The difference between $\hat{s}_{Y.X}$ and $\hat{s}\hat{y}$ can be easily recognized when plotted. The former are straight lines parallel to the regression line, whereas the latter are curved lines that are closest to the regression around the point $\bar{X}\ \bar{Y}$, and curve away from the regression line towards the ends of the distribution.

Example

Simple linear regression will be illustrated using the example adopted in Chapter 11, involving protein consumption in 1958 and incomes in 1968 in thirty-one countries. As will become apparent shortly, this particular example does not unequivocally meet the various assumptions and requirements of simple linear regression, hence it has the incidental advantage of illustrating some of the difficulties that geographers encounter in using simple regression.

To begin with, the designation of income as the dependent variable and protein consumption as the independent variable, as Belli (1971) did, may well be queried. Surely, it could be argued, the reverse is equally true. If the data were recorded for the same year, then it would probably make sense to reverse the positions of the variables, but there is a ten-year lag between the two variables: could income in 1968 influence protein consumption ten years earlier? The answer is a qualified no, which makes Belli's designation of protein consumption as the independent variable fairly reasonable. What the model is saying, therefore, is that a person's protein consumption influences his income-earning capacity in later years. Clearly, the model does not get at the process very closely, since there exists a complex system of mechanisms linking the two variables.

Referring to the seven assumptions of simple linear regression, the first is met since the data is measured at the ratio scale. Concerning the second, neither X (protein consumption) nor Y (income) is measured without error. On the other hand it is not difficult to make fairly accurate estimates of protein consumption whereas income is generally a less reliable measure. Hence strictly speaking the second assumption is not met, although the error in the measurement of X is probably sufficiently small that the model is still valid. Figure 11.10, plotting X against Y, shows that the third assumption of linearity is met: there is no evidence of a curvilinear relationship. The fourth assumption states that the errors, e, should not be spatially autocorrelated. The patchwork distribution of the countries makes it difficult to assess this, but an examination of the residuals for those pairs of countries that have common boundaries indicates that spatial autocorrelation amongst the regression residuals is largely absent.

Figure 13.6 shows the bivariate distribution of X and Y with the least squares regression line of Y on X plotted on it. A visual inspection of the conditional distribution of the errors provides no evidence to

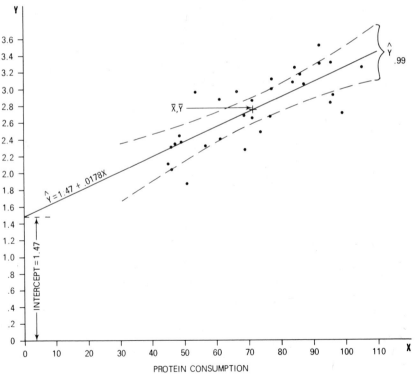

Figure 13.6: Bivariate distribution for X *and* Y *with regression line and 99% prediction interval for the regression line plotted.*

contradict the assumptions of zero means, constant variance, and normal distribution.

The seven assumptions are either satisfied, or at least are not grossly violated — with the possible exception of spatial autocorrelation in the *es*. This may lead to some loss of power in prediction of population parameters, but since the relationship being considered is fairly crude, it seems reasonable to proceed.

The intercept and slope coefficients will be calculated first so that a regression line may be fitted to the data. Using the column totals from Table 11.1, for equations **13.3** *and* **13.4** we have:

$$b_0 = \frac{(85 \cdot 151)\,(169940) - (2226)\,(6293 \cdot 9)}{(31)\,(169940) - 2226^2} = \frac{460340}{313064} = 1 \cdot 470$$

$$b_1 = \frac{(31)\,(6293 \cdot 9) - (2226)\,(85 \cdot 151)}{313064} = \frac{5564}{313064} = \cdot 0178$$

Hence we obtain the regression equation

$$\hat{Y} = 1 \cdot 470 + \cdot 0178\,X$$

Remember that a log transformation was applied to Y in Chapter 11, hence the regression, when expressed in terms of the original data, states that the log of income varies as a function of protein consumption.

A regression line is plotted on a graph by evaluating the equation for two values of X, one towards each end of the graph. In this case the equation was evaluated for $X = 0$ and $X = 100$, yielding

$$\hat{Y} = 1 \cdot 470 + (\cdot 0178)\,(0) = 1 \cdot 470$$

and $$\hat{Y} = 1 \cdot 470 + (\cdot 0178)\,(100) = 1 \cdot 470 + 1 \cdot 78 = 3 \cdot 25$$

A straight line is then drawn through the points $(0, 1 \cdot 470)$ and $(100, 3 \cdot 25)$. There is a simple check on the accuracy of the regression plot: the regression line should pass through the point $\bar{X}$, $\bar{Y}$, which is shown in Figure 13.6 ($\bar{X} = 71 \cdot 81$ and $\bar{Y} = 2 \cdot 75$).

In order to evaluate the formulae for the various standard errors, we require the calculations set out in Table 13.1, giving the sums of $(Y - \hat{Y})^2$ and $(\hat{Y} - \bar{Y})^2$.

Calculating the explained variance, $\hat{s}_E^2$ from **13.5**

$$\hat{s}_E^2 = \frac{3 \cdot 1902}{29} = \cdot 110 \text{ and } \hat{s}_E = \cdot 332$$

The variance of the regression residuals, according to **13.6** is

		Y	$\hat{Y}$	$Y-\hat{Y}$	$(Y-\hat{Y})^2$	$\hat{Y}-\bar{Y}$	$(\hat{Y}-\bar{Y})^2$
1	Argentina	2·715	3·212	–·497	·2470	·465	·2162
2	Austria	3·043	3·017	·026	·0007	·270	·0729
3	Brazil	2·401	2·555	–·154	·0237	–·192	·0369
4	Canada	3·320	3·159	·161	·0259	·412	·1697
5	Chile	2·667	2·839	–·172	·0296	·092	·0085
6	Columbia	2·418	2·324	·094	·0088	–·423	·1789
7	Denmark	3·289	3·106	·183	·0335	·359	·1289
8	Greece	2·814	3·159	–·345	·1190	·412	·1697
9	India	1·863	2·377	–·514	·2642	–·370	·1369
10	Ireland	2·924	3·177	–·253	·0640	·430	·1849
11	Israel	3·064	2·946	·118	·0139	·199	·0396
12	Italy	3·009	2·839	·170	·0289	·092	·0085
13	Japan	2·968	2·661	·307	·0942	–·086	·0074
14	Libya	2·967	2·413	·554	·3069	–·334	·1116
15	Mexico	2·679	2·679	·000	·0000	–·068	·0046
16	Netherlands	3·166	2·839	·327	·1069	·092	·0085
17	New Zealand	3·234	3·337	–·103	·0106	·590	·3481
18	Norway	3·223	2·964	·259	·0671	·217	·0471
19	Pakistan	2·033	2·288	–·255	·0650	–·459	·2107
20	Paraguay	2·283	2·679	–·396	·1568	–·068	·0046
21	Peru	2·382	2·342	·040	·0016	–·405	·1640
22	Philippines	2·367	2·306	·061	·0037	–·441	·1945
23	Portugal	2·626	2·732	–·106	·0112	–·015	·0002
24	Spain	2·849	2·732	·117	·0137	–·015	·0002
25	Sri Lanka	2·124	2·270	–·146	·0213	–·477	·2275
26	Syria	2·307	2·288	·019	·0004	–·459	·2107
27	Taiwan	2·344	2·484	–·140	·0196	–·263	·0692
28	Turkey	2·476	2·768	–·292	·0853	·021	·0004
29	Venezuela	2·881	2·555	·326	·1063	–·192	·0369
30	United Kingdom	3·193	2·999	·194	·0376	·252	·0635
31	United States	3·519	3·106	·413	·1706	·359	·1289
	Totals	85·151	85·152	–·004	2·1380	–·005	3·1902

$$s_{Y.X}^2 = \frac{2 \cdot 1380}{29} = \cdot 074 \text{ and } s_{Y.X} = \cdot 272$$

Since the total variance for Y is the sum of the explained and un-explained variance,

$$s_Y^2 = \frac{3 \cdot 1902}{29} + \frac{2 \cdot 1380}{29} = \frac{5 \cdot 3282}{29} = \cdot 184$$

and r, according to **13.7** is

$$r = \pm \sqrt{\frac{3 \cdot 1902}{5 \cdot 3282}} = \pm \sqrt{\cdot 5987} = \pm \cdot 774$$

Since this value of r agrees exactly with the value of r calculated in Chapter 11, we can be fairly sure that no serious arithmetic errors have been made up to this point.

Having calculated $s_{Y.X}$, the standard errors of b_0 and b_1 can be obtained from **13.8** and **13.9**, respectively. A table similar to Table 13.1 was used to calculate the sum of $(X - \bar{X})^2$, which is 10097·7.

$$s_{b_0} = \cdot 272 \sqrt{\frac{1}{31} + \frac{71 \cdot 806^2}{10097 \cdot 7}}$$

$$= \cdot 272 \sqrt{\cdot 0323 + \cdot 5106} = (\cdot 272)(\cdot 737)$$

$$= \cdot 200$$

$$s_{b_1} = \frac{\cdot 272}{\sqrt{10097 \cdot 7}} = \frac{\cdot 272}{100 \cdot 5} = \cdot 0027$$

Informally testing whether b_i is significantly positive ($H_1 : \beta_1 > 0$), we evaluate **13.11** and

$$t = \frac{b_1 - 0}{s_{b_1}} = \frac{\cdot 0178}{\cdot 0027} = 6 \cdot 60$$

As $t_{.01}$ in a one-tailed test with $v = 29$ is 2·46, we can conclude that the sample is drawn from a population in which the slope is significantly positive.

The final step is to place a confidence interval about the regression line. Since curved lines have to be plotted on the graph, $s_{\hat{y}}$ has to be evaluated for many values of X. The workings will not be presented in full, but by way of illustration, for the value $X = 50$, **13.12** becomes

$$\hat{s}\hat{y} = \cdot272 \sqrt{\frac{1}{31} + \frac{(50 - 71\cdot8)^2}{10097\cdot7}}$$

$$= \cdot272 \sqrt{\cdot0323 + \frac{475\cdot24}{10097\cdot7}}$$

$$= \cdot272 \sqrt{\cdot0323 + \cdot0471} = (\cdot272)(\cdot282) = \cdot0767$$

In order to plot the 99% prediction limit of the regression line, $\hat{s}\hat{y}$ is multiplied by $t_{\cdot01}$ (corresponding to a two-tailed test) with $N - 2$ degrees of freedom. Since $t_{\cdot01}$ ($v = 29$) = $2\cdot76$, for $X = 50$, the prediction interval

$$\hat{Y}_{\cdot99} = \pm (2\cdot76)(\cdot0767) = \pm \cdot212$$

The prediction intervals for the regression line are plotted in Figure 13.6. Characteristically, they are closest to the regression around $\bar{X}$, $\bar{Y}$, and curve away from it towards the extremes of the distribution.

What this prediction interval tells us is that 99 out of 100 times the mean value of Y corresponding to X for samples drawn from this population would fall within these limits. In consequence, those residuals that fall outside this confidence interval are ones in which we are particularly interested since in those countries, actual incomes are either higher, or lower than the 99% confidence limits of the regression line. The six countries with incomes that stand clearly above the 99% prediction interval are (from left to right on Figure 13.6): Libya; Venezuela; Japan; Netherlands; Norway; United States. Six countries fall below the prediction interval: India; Paraguay; Turkey; Greece; Ireland; Argentina. Libya, Venezuela, Norway and the United States have excellent resource endowments which contribute to income growth over and above the effect of nutrition. Japan and the Netherlands have higher than expected incomes due, perhaps, to the high level of entrepreneurship in these countries. Some of the inadequacies of the model, as formulated, are revealed by the negative residuals. For Ireland and Argentina, one may be tempted to argue that protein consumption is higher than incomes would lead one to expect due to the large local livestock industries. However such an argument contradicts the model which states that high protein consumption ought to be converted into high incomes in subsequent years. In the case of India and Paraguay, high population growth may be the cause of the negative residuals.

Geographical applications

Many of the applications of regression in geography use the model for descriptive rather than for inferential purposes. A good instance of this is Haggett's (1964) study of forest cover in southeast Brazil: the percentage of forest cover in his sample areas was regressed with:

1. Terrain index (a measure of relief)   +
2. Settlement spacing index   +
3. Rural population density index   −
4. Forest density index   +
5. Land values index   −

The + and − signs indicate whether the slope of the regression line was positive or negative (in each case the slope was in the expected direction).

Regressions of this type abound in the literature, and in student dissertations. The variable that is most commonly used as an independent variable is distance, for the very good reason that several of the theories and models used by geographers require a variable to change as a function of distance from a given node. Distance-dependent relationships are specified in both agricultural and urban land use theory and in diffusion and interaction models. Helvig's (1964) study of truck movements around Chicago is a good illustration of the latter since, when the population of a destination is held constant, the flow of trucks systematically decreases with increasing distance from Chicago. From one point of view, distance has the advantage of being a totally independent variable: distances are fixed quantities so that there cannot be a feedback loop from the dependent variable causing distances to change. From another point of view distance is an unsatisfactory independent variable since it is entirely passive in nature. Just as Rostow's *Stages of Economic Growth* has been criticized because he improperly characterizes time as a forcing function, so some geographers have incorrectly given distance an active, and therefore an 'explanatory' role in their regressions. In reality distance is a surrogate measure representing a spatial process.

Amongst the more interesting uses of simple regression by geographers are studies in which a functional interpretation can be placed on both of the regression coefficients. One such study that illustrates this use is Ambrose's (1970) analysis of patterns of growth in the Canadian labour force from 1951 to 1961. Ambrose fitted a model as follows:

$$\Delta E_S = b_0 + b_1 \, \Delta E_M$$

where ΔE_S is the change in employment in services from 1951 to 1961.

ΔE_M is the corresponding change in manufacturing employment. Since manufacturing employment is the independent variable, the model states that growth in service employment is some constant (b_0) plus a ratio (b_1) of the change in manufacturing employment. It follows that b_1 is a crude employment multiplier. For instance in the eighty major urban centres in Canada, Ambrose computed b_1 to be 3·42 meaning that for every new manufacturing job created from 1951 to 1961, nearly 3½ jobs were added in the service sector. The intercept, b_0, is a constant representing the change in service employment presumably attributable to the change in service demands of the population already living in a town in 1951. Thus each town in Canada was predicted to add 4459 new jobs in the service sector over and above those attributable to changes in manufacturing employment.

Colenutt's (1968) paper on linear predictive models in urban planning deals with a number of the difficulties and assumptions discussed earlier. He also refers to some of the simple manipulations which convert non-linear models to a linear form so that their parameters may be estimated by least squares. For instance, suppose that the number of trips (T) declines as a negative exponent of distance (D), hence

$$T = b_0 e^{-b_1 D}$$

where e (2·71828 . . .) is the base of the natural logarithms. Then applying natural logs, this relationship becomes linear and

$$\ln(T) = \ln b_0 - b_1 D$$

The S-shaped logistic curve, which is commonly used in diffusion studies to describe the number of people who have adopted an innovation, is another non-linear relationship that may be converted to a linear form. One version is as follows: given the logistic curve

$$Y_T = \frac{1}{1 + e^{b_0 - b_1 T}}$$

where Y_T is the percentage of the potential adopters who have actually adopted an innovation at time T, then simple manipulation gives:

$$e^{b_0 - b_1 T} = \frac{1 - Y_T}{Y_T}$$

taking natural logs,

$$\ln \left(\frac{1 - Y_T}{Y_T} \right) = b_0 - b_1 T$$

and b_0 and b_1 may be estimated by least squares.

At the suggestion of Thomas (1968) geographers have also used the regression model as an heuristic device by analysing regression residuals. A simple model is fitted and the residuals mapped: hopefully the mapped residuals will suggest further explanatory variables which are then included in a multiple regression model. A new set of residuals is mapped, and this sequence is repeated until the residuals appear to be randomly distributed. As an exploratory technique this is useful, although there is need for caution in drawing inferences, especially when the residuals display a high level of contiguity.

In conclusion, it is advisable to add a cautionary word about the use of regression in spatial analysis. Curry (1966, p. 97) went so far as to state that 'we still do not really know what we are doing in spatial regressions'. Since 1966 our ignorance about spatial regression has been reduced somewhat, particularly with respect to two aspects reviewed in the final chapter, namely ecological regressions and spatial auto-correlation. Nevertheless, it has to be admitted that our regression models are generally poor replicas of the actual processes involved; only rarely are the variables designated as independent, truly so, and there is often a conspicuous lack of theoretical underpinnings to our choice of variables.

Further readings in parametric statistics

The number of statistics textbooks is very large indeed: the following are recommended for their lucidity, and because they are written at the right level for students who have covered the material in this text. Blalock (1972), Croxton, Cowden and Klein (1968), Huntsberger and Billingsley (1973), Walker and Lev (1969), Yeates (1974), Yule and Kendall (1965).

14 Geography and statistical methods

Statistics is a science which ought to be honourable, the basis of many most important sciences; but it is not to be carried on by steam, this science, any more than any others are; a wise head is requisite for carrying it on. Conclusive facts are inseparable from inconclusive facts except by a head that already understands and knows.

THOMAS CARLYLE: *Chartism* II (1839)

Carlyle's penetrating remarks provide a fitting opening to this final chapter, since the aim is to discourage the mindless use of statistics in geography, whether by steam or by computer. This will be done in two ways: first, by setting in a broader context some of the techniques dealt with in the preceding chapters; and second, by considering three problem areas (namely closed number systems, ecological correlations, and spatial autocorrelation), which geographers commonly encounter in applying inferential statistics.

Although several of the techniques discussed in earlier chapters are not in common use by geographers, they were included because they provide useful introductions to more complex techniques. Five such 'stepping stones' were included.

1. The one-sample Kolmogorov—Smirnov test was used in Chapter 6 to compare an observed point pattern to a random point pattern with expected frequencies described by the Poisson distribution. This is the most basic point pattern test and leads into nearest neighbour techniques and into quadrat techniques (these are well reviewed in King, 1969). Point patterns are of considerable importance to geographers since they are an expression of the processes governing the spatial arrangement of phenomena.

2. The map comparison test developed by Court will probably not be extensively used by geographers, yet it introduces one of the most important topics in spatial analysis. Visualize map distributions as continuous surfaces with humps and troughs on them. Each surface has a characteristic spectrum which reflects the variance associated with harmonics (waves) of any particular wave length and amplitude. This is not an easy concept to grasp, but suppose, to take a simple physical example, that we have a surface recording heights in a large drumlin field with the drumlins spaced about half a mile apart: then the spectrum of that surface shows a peak at around half a mile because much of the variance of that surface is accounted for by harmonics with a wavelength

of about half a mile and an amplitude corresponding with the height of the drumlins (say 100 feet). Suppose that we also have a second surface measuring land use intensity over the drumlin field: the spectrum of this second surface may have two peaks, one at half a mile due to the effect of the drumlins, and another at five miles due to the effect of central places which in this area, we will assume, tend to be five miles apart. Ideally the topographic surface and the land-use intensity surface would be compared using a technique known as cross-spectral analysis, and one would obtain a measure known as the *coherence* of the two surfaces. However when geographers compare spatial distributions they generally fall back on correlation methods. Typically, our data is measured over districts, census tracts, square kilometres or whatever. Each such areal unit acts as a filter which eliminates variations whose wavelength is shorter in scale than the size of the filter. Thus if a half-mile square grid were imposed on the drumlin field and the average height in each grid square computed, then one would filter out most of the height variations due to the drumlins and end up with an apparently fairly uniform plane. This is where Court's test, or any correlation between surfaces, comes in since one is measuring the coherence of the surfaces at the scale of the areal units employed. If the phenomena being measured have entirely homogeneous distributions then the correlation (coherence) between the surfaces should vary in a predictable manner depending upon the level of aggregation. However if the distributions are not homogeneous, then the correlation will change irregularly depending upon the size of the filter used.

3. Dacey's contiguity measure provides a simple test of spatial autocorrelation. In a primitive way it begins to map the spectrum of a single variable and therefore may give some indication of the processes, contagious or otherwise, which govern spatial distributions.

4. From simple regression it is a small conceptual step to multiple regression. Computationally this is a much longer step since it is easiest to solve multiple regression problems using matrix algebra by phrasing the model as a version of the *general linear model*. Likewise simple correlation leads into a number of multivariate factorial methods based on regularities in the variance-covariance structure of a set of variables. Geographical versions of these multivariate methods are thoroughly treated in King (1969) and Yeates (1974).

5. McQuitty's linkage analysis is of interest because it is an elementary grouping algorithm which can be readily converted into a technique for regionalizing. This leads into the very large topic of taxonomy, and the geographical variation of this known as regional taxonomy. A review

article by Spence and Taylor (1970) reveals how large this field has grown.

These, then, are some of the statistical sub-fields into which the topics discussed in this book lead. The path, however, is often a stony one, and there have been several references to the difficulties that lie in the way of the venturous geographer. Three particular problem areas will now be briefly reviewed.

Percentages and closed number systems

Closed number systems are rarely referred to in the geographical literature, yet geographers make frequent use of this type of data. The most commonly encountered closed number system is data expressed as percentages. Consider the data in Table 14.1: notice that each row, including the column means, sum to 100%. Given the values in any two columns, the value in the third is fixed in order that the total sums to 100%.

Table 14.1: Hypothetical employment data illustrating a closed number system

| | Employment by industry type (%) | | | |
	Primary	Manufacturing	Service	Total
Town A	5	25	70	100
Town B	5	45	50	100
Town C	10	30	60	100
Town D	40	20	40	100
Arithmetic mean	15	30	55	100

Percentages have three undesirable properties that deserve mention. First, simple linear regressions fitted to such data can often lead to infeasible predictions — values under 0% and over 100%. Second, correlations between closed data variables should be treated with care: Krumbein (1962) shows how correlations based on open number systems can be very different from correlations based on the same data expressed as percentages. In a three component system such as Table 14.1 the relations amongst the variables are fixed in such a way that given the three sample variances, the correlation between any two variables can be predicted without measuring the covariance. Furthermore, there will always be two negative and one positive correlation when the variables of a three component system are correlated with each other, regardless of the signs attached to the correlations in the raw data (Chayes, 1971). The third undesirable property relates to binomially distributed variables, which is what dichotomized percentage variables are in practice. Recall that in Chapter 5, the binomial distribution was used to describe situations with two outcomes. The example of people with geography degrees is precisely such a situation: there are two outcomes, people with, and people without geography degrees. We know that 3·2% of the population have geography degrees, but this cannot be applied to an individual who either has, or does not have a degree: hence we can say that an individual has a probability, P, of ·032 of having a degree in geography but we cannot say that the average person has ·032 of a geography degree. Wrigley (1973) shows that a binomially distributed dependent variable will violate the assumption of homoscedasticity in the simple linear regression model.

How are geographers to handle percentage data? The best answers available to date have been provided by Wrigley, who recommends a logit (short for logistic) transformation similar to that outlined at the end of the previous chapter. If PC_j is the percentage of the people in the jth areal unit who have geography degrees, or whatever, then the logit transformation

$$L_j = \ln \left(\frac{PC_j}{100 - PC_j} \right) \qquad 14.1$$

produces a curve that asymptotically approaches 0% and 100%, so that infeasible predictions are eliminated. Concerning the problem that a binomially distributed variable does not have constant error variance, Wrigley suggests that ordinary least squares estimates be replaced by weighted least squares — a topic beyond the scope of this book.

Ecological correlation and modifiable units

When the phi coefficient was discussed in Chapter 7, reference was made to W. S. Robinson (1950) who found that the correlation between two variables changed greatly when different areal aggregations of the same data were used. We can apply the term 'ecological' to any study using areal units as observations. Hence Robinson's study makes use of ecological correlations as do several of the examples presented earlier in this book including Spearman's rank correlation of El Salvador data (Chapter 7) and Pearson's r applied to protein consumption and economic development (Chapter 11). Indeed the majority of correlations computed by geographers are ecological correlations.

Modifiable units were referred to in dealing with geographical examples of the chi square test. Any arbitrary unit of measurement — lengths, areas, weights, time periods, etc. — are modifiable so that statistical measures computed for, say, square miles are likely to be different from the same measures based on square kilometre recording units. It should be clear, then, that ecological correlations are a special case of the more general problem of modifiable units.

The problem of ecological correlations has been known of for some time, and has received considerable attention, particularly on the part of sociologists: the large volume edited by Dogan and Rokkan (1969) contains a number of essays that deal with the problem exhaustively.

What has to be addressed is the question: 'What meaning should a geographer attach to an ecological correlation?' Perhaps the most important thing to stress is that inferences should be made strictly in terms of the areal units that are employed. Thus if the correlation between age and income measured over sixty-five census tracts is ·5, then this relationship applies to that particular sampling frame. It would be quite incorrect, without a lot of other information, to extrapolate this result and say that the same correlation of ·5 obtains for individual residents: the individual correlation may be quite different. Except under special circumstances it would also be incorrect to assert that the correlation of ·5 applies to some other scale of areal aggregation. Figure 14.1 illustrates why this is so. In A, sixteen individuals are plotted in a 4 x 4 matrix: each individual is recorded on 2 variables, X and Y, with values of either +1 or −1. The upper value in each box is X, the lower, Y. The scatter diagram shown in B corresponds with the data in A: it is clear that no relationship between X and Y exists at the individual level and $r = 0$. In C the individuals are grouped into areas in such a way that there are two individuals in each area: the resulting scatter diagram for the eight areas (shown in D) reveals a perfect positive correlation and $r = +1$. Likewise in E, the individuals are grouped into 8 small areas containing two individuals, but the areas are different from C. In this case F reveals a perfect negative correlation and $r = −1$. The example is hypothetical but it has been demonstrated that ecological correlations of +1 and −1 have been obtained from a set of observations with an individual correlation of zero.

The discussion of Court's test earlier in this chapter gives some indication of what causes ecological correlations to behave the way they do. An ecological correlation compares two surfaces with distinctive

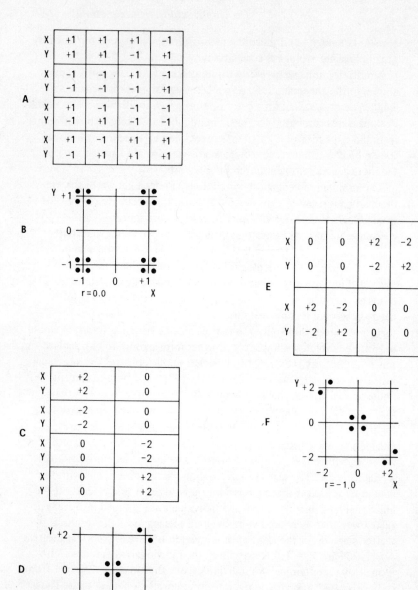

Figure 14.1: Individual and ecological correlations for a hypothetical example.

power spectra using a particular set of areal units that captures the co-variation of the surfaces for that particular scale of aggregation. If the distributions are homogeneous, then the greater the level of aggregation, the more will random fluctuations be ironed out and the stronger will be the correlation (either positive or negative) between the two variables: thus in Robinson's (1950) example, the individual correlation was $+.203$, for the forty-eight states $r = +.773$ and for the nine major regions $r = +.946$.

If inferences are made specifically for the set of areal units employed in the sampling frame, then ecological correlations present no special problems of interpretation. Furthermore, there are geographers who argue that since areas or regions are the object of geographical study, this is precisely what we should be doing. On the other hand there are geographers, particularly those favouring a behavioural approach, who wish to know about the behaviour of individuals, indeed if one wishes to identify the processes that give rise to spatial distributions, then a knowledge of the behaviour of individuals is often invaluable. This being the case, can statements about individuals be made on the basis of areally grouped data? The answer is a qualified yes. Duncan and Davis (1953) have suggested that by estimating the marginal totals of a 2 x 2 contingency table using areally grouped data, the individual correlation can at least be placed within bounds, and the finer the areal disaggregation, the narrower the bounds. An alternative approach, suggested by Goodman (1959) is to fit regression lines to the scatter diagrams for ecological data and then use the intercept and slope coefficients to estimate the probability of an individual belonging in any of the cells of a 2 x 2 table: this in turn is used to estimate the individual correlation.

Ecological correlations raise one other topic of geographical interest. Suppose an ecological correlation is to be computed for areas of unequal size: should the areas be weighted in some way so as to take size inequalities into account? A. H. Robinson (1956) advocates the use of weights, and has presented formulae for estimating correlation and regression coefficients with areas weighted according to their size. Curry (1966) has taken up this problem and concludes that weighting may be justifiable in certain circumstances, but in direct contrast to Robinson, he suggests that the smallest areas be given the biggest weighting! His argument relates to the power spectra of the two variables: in essence he suggests that the smallest areal units having the finest mesh will trap the greatest amount of information concerning the relationship between the two variables.

Spatial autocorrelation

An autocorrelated series is one which is related to itself, so that observations tend to be correlated with other observations that stand next to them in the series (either in time or in space). The concept is most easily explained for time series since they are one-dimensional: suppose that wet years tend to be followed by wet years, and dry years by dry years in such a way that they run in seven year cycles — seven years of drought and want followed by seven years of plenty. This would give rise to an autocorrelated time series. Likewise in space, although the matter becomes much more complicated since spatial series are two-dimensional.

Statistical theory for autocorrelated time series has been developed for a long time, but only in the last decade has the spatial version received much attention. Dacey's contiguity test discussed in Chapter 12 is perhaps the simplest test of spatial autocorrelation. Subsequently, in one of the most important co-operative research ventures in recent years, Cliff (a geographer) and Ord (an econometrician) have made major contributions to the statistical theory of spatial autocorrelation and autoregression. Much of their work is summarized in their joint book, *Spatial Autocorrelation* (1973). Amongst other things, Cliff and Ord have developed a new measure of spatial autocorrelation which is superior to earlier measures suggested by Geary and by Moran. The statistic will not be presented here since it leads into some quite advanced statistical theory.

The impression may have been given that spatial autocorrelation is a bit of a nuisance, since its absence appears as an assumption for linear regression. If that impression has been given it must be corrected: from a geographical point of view it is an extremely valuable effect which provides information on the processes that operate in space. If spatial autocorrelation is detected in a series, then immediately the geographer has a problem to solve since something must have caused the series to be autocorrelated: subject to Type I errors, the series cannot have been produced randomly.

In seeking to account for autocorrelation in spatial series, geographers can fall back on two broad types of explanation, one ecological or 'vertical', the other spatial or 'horizontal'. Suppose that the yield of maize on smallholdings in Kenya was being examined, and two broadly defined regions were found, one with high yields, the other with low yields. An ecological explanation might relate yields to local soils: the high yield region might be on volcanic soils, the low yield region on lateritic soils. Hence two adjoining townships might have high yields because of vertical man–land relationships within each township. A spatial explanation, on the other hand, might invoke the diffusion of hybrid corn or of the use of fertilizers as the cause of high yields so that areas late in adopting these new techniques would be low yield areas. Provided the diffusion of the new techniques is subject to a neighbourhood effect, the resulting spatial series should display an autocorrelated pattern.

Concluding remarks

This book is cast in an inferential framework in the belief that statistics have an important role to play in geographical analysis, and hence in developing geographical theory. Nevertheless, the use of statistics is not something to be blindly advocated. The assumptions and limitations of specific tests have been reviewed in earlier chapters. This concluding section turns to some of the issues underlying the overall validity of the inferential approach. A number of these issues have been identified by Olsen (1969) and by Gould (1970) in his controversial paper 'Is *statistix inferens* the geographical name for a wild goose?' in which he concludes (p. 447) with a very gloomy remark concerning 'the devastating weaknesses that make the general approach of inferentialism in geographical research so doubtful'. This author does not share Gould's gloomy viewpoint. Moreover the title of Gould's paper would suggest that his strictures on the use of statistical inference are more all embracing than they really are. In practice his remarks are addressed largely to the use of correlation and regression in geography, indeed a number of his specific criticisms concern multiple regression which has not been discussed in this book. Only a few of his concerns relate to the use of nonparametric methods.

Seven broad issues concerning the validity of the inferential approach will be considered by way of a dialogue: the issue will be posed, with the position adopted here as an answer.

Issue: The major value of a course in statistics is that it provides an intellectual exercise in clear thinking and problem solving, combined with a 'serendipity effect' (implying that in struggling with statistics a geographer will be confronted by all sorts of intriguing geographical problems).

Answer: While accepting the pedagogic value of statistics, it is felt that their main value lies in their role in the system of scientific explanation.

Issue: Geographers generally use the ·05 or ·01 significance level without considering why such values of α are appropriate.

Answer: This allegation is largely true, although the repeated and exhaustive defence of a particular α every time a test is conducted will soon prevent the wood from being seen for the trees. In Chapter 2 it was suggested that the 'cost' of making a mistake (particularly a Type I error) should be taken into account in selecting α.

Issue: Geographers often use statistical populations in their analyses, and then draw inferences as if they had worked with a sample drawn from a larger population.

Answer: This was also considered in Chapter 2: the discussion inclined to the argument that a hypothetical larger population can generally be defined — usually because of the influence of stochastic processes.

Issue: Linear regression requires that the relationship between two variables be linear, so geographers apply transformations in order to meet this assumption. The result, however, is a predictive equation that does not correspond with the non-linear functional relationship in the real world.

Answer: We can fit parabolas and other higher order functions to our data so as to avoid using transformations. However the resulting predictive equations are often harder to interpret than the equations based on transformed data. Besides, the vast majority of transformations are logarithmic in form, and it often makes sense to state that there is a functional relationship such that one variable varies as the log of another variable. The statistical convenience of linearity is not necessarily incompatible with the examination of non-linear relationships in the real world.

Issue: Despite the fact that the majority of geographical variables are patently non-normal, we squeeze them into the normal mould by transformations in order to take advantage of standard correlation and regression theory.

Answer: The pay-off between the empirical meaning lost in transforming data and the power gained in meeting the normality assumption often justifies the use of transformations. Besides, as was stated in Chapters 11 and 13, inferential methods for correlation and regression appear to be fairly insensitive to moderate violations of the normality assumption.

Issue: Many geographical variables are spatially autocorrelated, thereby violating the assumption of mutually independent observations.

Answer: This is a central problem, and has been discussed at some length in Chapter 13 and in this chapter.

Issue: Given the widespread availability of large computers, geographers should make greater use of randomization techniques rather than standard tests of inference to make probabilistic statements.

Answer: For those who pay their own computer bills 'running the problem a hundred or a thousand times' is not quite as attractive as it sounds. Running up a bill for a few hundred dollars every time one wants to plot the sampling distribution for N! random permutations can become a bit expensive. Besides, this is not necessary in many cases: for instance Norcliffe (1969) shows how critical values for certain trend surface models obtained by randomization methods correspond extremely closely with critical values given by the F distribution. This is not to dismiss the use of randomization but to caution against using expensive computer time when simple look-up tables achieve the same result, and also to recommend the use of randomization specifically to check how closely the various theoretical distributions approximate empirical sampling distributions when various assumptions are abused.

Where does the reader go next? Almost certainly, having mastered the basics, the next steps are into linear algebra since multivariate statistical methods can best be expressed in terms of matrix algebra. Yeates (1974) does this well. For students with an aptitude for mathematics, a course in mathematical statistics is strongly recommended. Beyond that there is the whole field of spatial analysis including many of the topics covered by King (1969). Whatever the route, let it be hoped that statistics will be used not just for their intrinsic interest but as a source of illumination as we strive for fuller and better explanations of those things that fall within the compass of geography.

Bibliography

ADAMS, J. S. (1969), Directional bias in intra-urban migration, *Economic Geography,* **45**, 302–23.

AMBROSE, P. (1970), Patterns of growth in the Canadian labour force 1951–1961, *Canadian Geographer,* **14**, 139–57.

AMEMIYA, E. C. (1964), Economic differentiation and social organization of standard metropolitan areas, *Journal of Regional Science,* **5**, 57–61.

ANDERSON, N. M. (1961), Scales and statistics: parametric and non-parametric, *Psychological Bulletin,* **58**, 305–16.

BACHI, R. (1957), Statistical analysis of geographical series, *Bulletin de l'Institut International de Statistique,* **36**, 229–40. Reprinted in *Spatial Analysis,* Berry, B. J. L., and Marble, D. F. (eds.), 101–9.

BARTLETT, M. S. (1947), The use of transformations, *Biometrics,* **3**, 39–52.

BATSCHELET, E. (1965), *Statistical Methods for the Analysis of Problems in Animal Orientation and Certain Biological Rhythms,* American Institute of Biological Sciences (Monograph Series).

BELLI, P. (1971), The economic implications of malnutrition: the dismal science revisited, *Economic Development and Cultural Change,* **20**, 1–23.

BERRY, B. J. L., and BAKER, A. M. (1968), Geographic sampling, in *Spatial Analysis,* Berry, B. J. L., and Marble D. F. (eds.), 91–100.

BERRY, B. J. L., and GARRISON, W. L. (1958), The functional bases of the central place hierarchy, *Economic Geography,* **34**, 145–54.

BERRY, B. J. L., and MARBLE, D. F. (1968) (eds.), *Spatial Analysis,* Englewood Cliffs: Prentice-Hall.

BLAIKIE, P. M. (1971), Spatial organization of agriculture in some North Indian villages: part I, *Transactions,* Institute of British Geographers, **52**, 1–40.

BLALOCK, H. M. (1972), *Social Statistics,* New York: McGraw-Hill.

CHAYES, F. (1971), *Ratio Correlation: A Manual for Students of Petrology and Geochemistry,* Chicago: University of Chicago Press.

CHOYNOWSKI, M. (1959), Maps based on probabilities, *Journal of the American Statistical Association,* **54**, 385–8. Reprinted in *Spatial Analysis,* Berry, B. J. L., and Marble, D. F. (eds.), 180–83.

CLARKE, R. D. (1948), An application of the Poisson distribution, *Journal of the Institute of Actuaries,* 72 (no. 335), 481.

CLIFF, A. D. (1967), Some measures of contiguity for two colour mosaic maps with vacancies, Department of Geography, University of Bristol, Discussion Paper Series A, **4**.

CLIFF, A. D. (1973), A note on statistical hypothesis testing, *Area,* **5**, 240.

CLIFF, A. D., and ORD, J. K. (1969), The problem of spatial auto-correlation, in *Studies in Regional Science,* Scott, A. J. (ed.), (London: Pion) 25–55.

CLIFF, A. D., and ORD, J. K. (1973), *Spatial Autocorrelation,* London: Pion.

COCHRAN, W. G. (1947), Some consequencies when the assumptions of analysis of variance are not satisfied, *Biometrics,* **3**, 22–38.

COLENUTT, R. J. (1968), Building linear predictive models for urban planning, *Regional Studies,* **2**, 139–43.

CONOVER, W. J. (1971), *Practical Nonparametrical Statistics,* New York: Wiley.

COURT, A. (1966), The inter-neighbour interval, *Yearbook,* Association of Pacific Coast Geographers, **28**, 180–82.

COURT, A. (1970), Map comparisons, *Economic Geography,* **46**, 435–8.

COURT, A. (1972), All statistical populations are estimated from samples (with reply by D. R. Meyer), *Professional Geographer,* **24**, 160–62.

CROXTON, F. E., COWDEN, D. J., and KLEIN, S. (1968 – 3rd edition), *Applied General Statistics,* London: Pitman.

CURRY, L. (1966), A note on spatial association, *Professional Geographer,* **18**, 97–9.

CURRY, L. (1967), Chance and Landscape, in *Northern Geographical Essays,* House, J. W. (ed.), Newcastle-upon-Tyne: Oriel Press.

DACEY, M. F. (1968), A review of measures of contiguity for two and *K*-color maps, in *Spatial Analysis,* Berry, B. J. L., and Marble, D. F. (eds.), 479–95.

DACEY, M. F. (1968), An empirical study of the areal distribution of houses in Puerto Rico, *Transactions,* Institute of British Geographers, **45**, 51–69.

DAVIS, J. T. (1971), Sources of variation in housing values in Washington DC, *Geographical Analysis,* **3**, 63–76.

DOGAN, M., and ROKKAN, S. (1969) (eds.), *Quantitative Ecological Analysis in the Social Sciences,* Cambridge, Mass.: MIT Press.

DONALDSON, T. S. (1968), Robustness of the *F*-test to errors of both kinds and the correlation between the numerator and denominator of the *F*-ratio, *Journal of the American Statistical Association,* **63**, 660–76. Reprinted in *Readings in Statistics for the Behavioural Sciences,* Heerman, E. F., and Braskamp, L. A. (eds.).

DOORNKAMP, J. C., and KING, C. A. M. (1971), *Numerical Analysis in Geomorphology,* London: Arnold.

DUNCAN, O. D., and DAVIS, B. (1953), An alternative to ecological correlation, *American Sociological Review,* **18**, 665–6.

EVANS, I. S. (1969), The geomorphology and morphometry of glacial and nival areas, in *Water, Earth and Man,* R. J. Chorley (ed.), (London: Methuen), 369–80.

EVANS, I. S. (1972), Inferring process from form: the asymmetry of glaciated mountains, *International Geography,* **1**, 17–19.

GAITO, J. (1959), Nonparametric methods in psychological research, *Psychological Reports,* **5**, 115–25.

GAITO, J. (1973), *Introduction to Analysis of Variance Procedures,* New York: MSS Information Corporation.

GETIS, A. (1964), Temporal land-use pattern analysis with the use of nearest neighbour and quadrat methods, *Annals,* Association of American Geographers, **54**, 391–9.

GOODMAN, L. A. (1954), Kolmogorov–Smirnov tests for psychological research, *Psychological Bulletin,* **51**, 160–68.

GOODMAN, L. A. (1959), Some alternatives to ecological correlation, *American Journal of Sociology,* **64**, 610–25. Reprinted in *Spatial Analysis,* Berry, B. J. L., and Marble, D. F. (eds.), 447–60.

GOULD, P. (1970), Is *statistix inferens* the geographical name for a wild goose?, *Economic Geography,* **46**, 439–48.

GOULD, P. R., and OLA, D. (1970), The perception of residential desirability in the western region of Nigeria, *Environment and Planning,* **2**, 73–87.

GREER-WOOTTEN, B. (1972), *A Bibliography of Statistical Applications in Geography,* Association of American Geographers, Commission on College Geography, Technical Paper No. 9.

GREER-WOOTTEN, B., and GILMOUR, G. M. (1972), Distance and directional bias in migration patterns in depreciating metropolitan areas, *Geographical Analysis,* **4**, 92–7.

HAGGETT, P. (1964), Regional and local components in the distribution of forested areas in southeast Brazil: a multivariate approach, *Geographical Journal,* **130**, 365–77.

HAGGETT, P. (1965), *Locational Analysis in Human Geography*, London: Arnold.

HAJEK, J. (1969), *A Course in Nonparametric Statistics*, San Francisco: Holden-Day.

HARTSHORNE, R. (1959), *Perspective on the Nature of Geography*, Chicago: Rand McNally.

HARVEY, D. W. (1966), Geographical processes and point patterns: testing models of diffusion by quadrat sampling, *Transactions*, Institute of British Geographers, **40**, 81–95.

HARVEY, D. W. (1969), *Explanation in Geography*, London: Arnold.

HAWORTH, J., and VINCENT, P. (1974), Calculation of prediction limits in linear regression, *Area*, **6**, 113–16.

HEERMANN, E. F., and BRASKAMP, L. A. (1970), *Readings in Statistics for the Behavioural Sciences*, Englewood Cliffs: Prentice-Hall.

HELVIG, M. (1964), *Chicago's External Truck Movements: Spatial Interaction between the Chicago Area and its Hinterland*, University of Chicago, Department of Geography, Research Paper No. 90.

HUNTSBERGER, D. V., and BILLINGSLEY, P. (1973 – 3rd edition), *Elements of Statistical Inference*, Boston: Allyn & Bacon.

JOHNSTON, R. J. (1970), On spatial patterns in the residential structure of cities, *Canadian Geographer*, **14**, 361–7.

KENDALL, M. G. (1962 – 3rd edition), *Rank Correlation Methods*, London: Griffin.

KING, L. J. (1961), A multivariate analysis of the spacing of urban settlements in the United States, *Annals*, Association of American Geographers, **51**, 222–33.

KING, L. J. (1969), *Statistical Analysis in Geography*, Englewood Cliffs: Prentice-Hall.

KNOS, D. S. (1962), *Distribution of Land Values in Topeka, Kansas*, Lawrence, Kansas: Bureau of Business and Economic Research. Reprinted in *Spatial Analysis*, Berry, B. J. L., and Marble, D. F. (eds.), 269–89.

KRUMBEIN, W. C. (1939), Preferred orientation of pebbles in sedimentary deposits, *Journal of Geology*, **47**, 673–706.

KRUMBEIN, W. C. (1962), Open and closed number systems in stratigraphic mapping, *Bulletin of the American Association of Petroleum Geologists*, **46**, 2229–45.

KRUSKAL, W. H. (1958), Ordinal measures of association, *Journal of the American Statistical Association*, **53**, 814–61.

KUH, E., and MEYER, J. R. (1955), Correlation and regression estimates when the data are ratios, *Econometrica,* **23**, 400–16.

McCARTY, H. H., HOOK, J. C., and KNOS, D. S. (1956), *The Measurement of Association in Industrial Geography,* Iowa City: State University of Iowa Press.

MACKAY, J. R. (1958), Chi-square as a tool for regional studies, *Annals,* Association of American Geographers, **58**, 164.

McQUITTY, L. L. (1957), Elementary linkage analysis for isolating orthogonal and oblique types and typal relevancies, *Educational and Psychological Measurement,* **17**, 207–29.

MARTIN, R. L. (1974), On spatial dependence, bias and the use of first spatial differences in regression analysis, *Area,* **6**, 185–94.

MEYER, D. R. (1972), Geographical population data: statistical description not statistical inference, *Professional Geographer,* **24**, 26–8.

MORONEY, M. J. (1956), *Facts from Figures,* Harmondsworth: Penguin.

MORRILL, R. L., and PITTS, F. R. (1967), Marriage, migration and the mean information field: a study in uniqueness and generality, *Annals,* Association of American Geographers, **57**, 401–22.

MURDIE, R. A. (1969), *Factional Ecology of Metropolitan Toronto 1951–1961,* Department of Geography, University of Chicago, Research Paper No. 116.

MURDIE, R. A. (1970), A reply to 'On spatial patterns in the residential structure of cities', *Canadian Geographer,* **14**, 367–9.

NEFT, D. S. (1966), *Statistical Analysis for Areal Distributions,* Regional Science Research Institute, Monograph Series No. 2.

NEYMAN, J. (1953), *First Course in Probability and Statistics,* New York: Holt.

NORCLIFFE, G. B. (1968), Areal grouping with elementary and nearest neighbour linkage analysis, *Department of Geography, University of Bristol,* Seminar Paper Series A, 12.

NORCLIFFE, G. B. (1969), On the use and limitations of trend surface models, *Canadian Geographer,* **13**, 338–48.

NORCLIFFE, G. B. (1974), Territorial influences in urban political space: a study of perception in Kitchener–Waterloo, *Canadian Geographer,* **18**, 311–29.

NORCLIFFE, G. B., and HEIDENREICH, C. E. (1974), The preferred orientation of Iroquoian longhouses in Ontario, *Ontario Archeology,* **23**, 3–30.

NORCLIFFE, G. B., and MITCHELL, P. (n.d.), Regional productivity variations in Canada, *in preparation*.

NYSTUEN, J. D. (1963), Identification of some fundamental spatial concepts, *Papers of the Michigan Academy of Science, Arts and Letters*, **48**, 373–84. Reprinted in *Spatial Analysis*, Berry, B. J. L., and Marble, D. F. (eds.), 35–41.

OLSSON, G. (1969), Inference problems in locational analysis, *Behavioural Models in Geography: A Symposium. Northwestern University Studies in Geography, No. 17*, Cox, K. R., and Golledge, R. G. (eds.), 14–33.

PAPAGEORGIOU, G. J. (1969), Description of a basis necessary to the analysis of spatial systems, *Geographical Analysis*, **1**, 213–15.

PARASKEVOPOULOS, C. C. (1971), The stability of the regional share component: an empirical test, *Journal of Regional Science*, **11**, 107–12.

PIERCE, A. (1970), *Fundamentals of Nonparametric Statistics*, Belmont, California: Dickenson.

POOLE, M. A., and O'FARRELL, P. N. (1971), The assumptions of the linear regression model, *Transactions*, Institute of British Geographers, **52**, 145–58.

QUENOUILLE, M. H. (1952), *Associated Measurements*, London: Butterworth.

RICHTER, C. E. (1969), The impact of industrial linkages on geographical association, *Journal of Regional Science*, **9**, 19–28.

ROBINSON, A. H. (1956), The necessity of weighting values in correlation analysis of areal data, *Annals*, Association of American Geographers, **46**, 233–6.

ROBINSON, G., and FAIRBAIRN, K. J. (1969), An application of trend surface mapping to the distribution of residuals from a regression, *Annals*, Association of American Geographers, **59**, 158–70.

ROBINSON, W. S. (1950), Ecological correlations and the behaviour of individuals, *American Sociological Review*, **15**, 351–7.

ROZEBOOM, W. H. (1960), The fallacy of the null-hypothesis significance test, *Psychological Bulletin*, **57**, 416–28. Reprinted in *Readings in Statistics for the Behavioural Sciences*, Heerman, E. F., and Braskamp, L. A. (eds.).

RUSHTON, G., GOLLEDGE, R. G., and CLARK, W. A. V. (1967), Formulation and test of a normative model for the allocation of grocery expenditures by a dispersed population, *Annals*, Association of American Geographers, **57**, 389–400.

SAWREY, E. (1958), A distinction between exact and approximate nonparametric statistics, *Psychometrika,* **23**, 171–7. Reprinted in *Readings in Statistics for the Behavioural Sciences,* Heermann, E. F., and Braskamp, L. A. (eds.).

SCHAEFER, F. K. (1953), Exceptionalism in geography: a methodological examination, *Annals,* Association of American Geographers, **43**, 226–49.

SIEGEL, S. (1956), *Nonparametric Statistics for the Behavioural Sciences,* New York: McGraw-Hill.

SMITH, W. (1955), The location of industry, *Transactions,* Institute of British Geographers, **21**, 1–18.

SOKAL, R. R., and SNEATH, P. H. A. (1963), *Principles of Numerical Taxonomy,* San Francisco: Freeman.

SPENCE, N. A., and TAYLOR, P. J. (1970), Quantitative methods in regional taxonomy, in Board, C., *et al.* (eds.), *Progress in Geography,* **2**, 1–63.

STAFFORD, H. A. (1966), Population as a determinate of industrial type, *East Lakes Geographer,* **2**, 71–9.

STEVENS, S. S. (1946), On the theory of scales of measurement, *Science,* **103**, 677–80.

TAAFFE, E. J., MORRILL, R. L., and GOULD, P. R. (1963), Transport expansion in underdeveloped countries: a comparative analysis, *Geographical Review,* **53**, 503–29.

THOMAS, E. N. (1968), Maps of residuals from regression: their characteristics and uses in geographic research, in Berry, B. J. L., and Marble, D. F. (eds.), *Spatial Analysis,* 326–52.

TUKEY, J. W. (1962), The future of data analysis, *Annals of Mathematical Statistics,* **33**, 1–67.

UNWIN, D. J. (1973), The distribution and orientation of corries in northern Snowdonia, *Transactions,* Institute of British Geographers, **58**, 85–97.

WALKER, H. M., and LEV, J. (1953), *Statistical Inference,* New York: Holt, Rinehart & Winston.

WALKER, H. M., and LEV, J. (1969 – 3rd edition), *Elementary Statistical Methods,* New York: Holt, Rinehart & Winston.

WHITE, R. R. (1970), *Geographic Information and the Interpretation of Choropleth Maps,* Unpublished Ph.D. dissertation, Bristol University.

WHITE, R. R. (1971), Probability maps of leukemia mortalities in England and Wales, in *Readings in Medical Geography,* McGlashan, N. (ed.), London: Methuen.

WOLPERT, J. (1967), Distance and directional bias in inter-urban migratory streams, *Annals,* Association of American Geographers, **57**, 605–16.

WRIGLEY, N. (1973), The use of percentages in geographical research, *Area,* **5**, 183–6.

YEATES, M. (1974), *An Introduction to Quantitative Analysis in Human Geography,* New York: McGraw-Hill.

YULE, G. U., and KENDALL, M. G. (1965 – 14th edition), *An Introduction to the Theory of Statistics,* London: Griffin.

ZAIDI, I. H. (1968), Measuring the locational complementarity of central places in West Pakistan: a macrogeographic framework. *Economic Geography,* **44**, 218–39.

ZOBLER, L. (1958), The distinction between relative and absolute frequencies in using chi-square for regional analysis, *Annals,* Association of American Geographers, **48**, 456–7.

Statistical tables

Table A: Values of $e^{-\lambda}$ for the Poisson distribution

$(0 < \lambda < 1)$

λ	·00	·01	·02	·03	·04	·05	·06	·07	·08	·09
0·0	1·0000	·9900	·9802	·9704	·9608	·9512	·9418	·9324	·9231	·9139
0·1	·9048	·8958	·8869	·8781	·8694	·8607	·8521	·8437	·8353	·8270
0·2	·8187	·8106	·8025	·7945	·7866	·7788	·7711	·7634	·7558	·7483
0·3	·7408	·7334	·7261	·7189	·7118	·7047	·6977	·6907	·6839	·6771
0·4	·6703	·6636	·6570	·6505	·6440	·6376	·6313	·6250	·6188	·6126
0·5	·6065	·6005	·5945	·5886	·5827	·5770	·5712	·5655	·5599	·5543
0·6	·5488	·5434	·5379	·5326	·5273	·5220	·5169	·5117	·5066	·5016
0·7	·4966	·4916	·4868	·4819	·4771	·4724	·4677	·4630	·4584	·4538
0·8	·4493	·4449	·4404	·4360	·4317	·4274	·4232	·4190	·4148	·4107
0·9	·4066	·4025	·3985	·3946	·3906	·3867	·3829	·3791	·3753	·3716

$(\lambda = 1, 2, 3, \ldots, 10)$

λ	1	2	3	4	5	6	7	8	9	10
$e^{-\lambda}$	·36788	·13534	·04979	·01832	·006738	·002479	·000912	·000335	·000123	·000045

To calculate values of $e^{-\lambda}$ for other values of λ use the law of exponents. For instance,
$e^{-1·55} = (e^{-1·00})(e^{-0·55}) = (·36788)(·5770) = ·2123$.

Table B: Critical values for chi square

ν						Level of significance (α)								
	·99	·98	·95	·90	·80	·70	·50	·30	·20	·10	·05	·02	·01	·0
1	·00016	·00063	·0039	·016	·064	·15	·46	1·07	1·64	2·71	3·84	5·41	6·64	10·
2	·02	·04	·10	·21	·45	·71	1·39	2·41	3·22	4·60	5·99	7·82	9·21	13·
3	·12	·18	·35	·58	1·00	1·42	2·37	3·66	4·64	6·25	7·82	9·84	11·34	16·
4	·30	·43	·71	1·06	1·65	2·20	3·36	4·88	5·99	7·78	9·49	11·67	13·28	18·
5	·55	·75	1·14	1·61	2·34	3·00	4·35	6·06	7·29	9·24	11·07	13·39	15·09	20·
6	·87	1·13	1·64	2·20	3·07	3·83	5·35	7·23	8·56	10·64	12·59	15·03	16·81	22·
7	1·24	1·56	2·17	2·83	3·82	4·67	6·35	8·38	9·80	12·02	14·07	16·62	18·48	24·
8	1·65	2·03	2·73	3·49	4·59	5·53	7·34	9·52	11·03	13·36	15·51	18·17	20·09	26·
9	2·09	2·53	3·32	4·17	5·38	6·39	8·34	10·66	12·24	14·68	16·92	19·68	21·67	27·
10	2·56	3·06	3·94	4·86	6·18	7·27	9·34	11·78	13·44	15·99	18·31	21·16	23·21	29·
11	3·05	3·61	4·58	5·58	6·99	8·15	10·34	12·90	14·63	17·28	19·68	22·62	24·72	31·
12	3·57	4·18	5·23	6·30	7·81	9·03	11·34	14·01	15·81	18·55	21·03	24·05	26·22	32·
13	4·11	4·76	5·89	7·04	8·63	9·93	12·34	15·12	16·98	19·81	22·36	25·47	27·69	34·
14	4·66	5·37	6·57	7·79	9·47	10·82	13·34	16·22	18·15	21·06	23·68	26·87	29·14	36·
15	5·23	5·98	7·26	8·55	10·31	11·72	14·34	17·32	19·31	22·31	25·00	28·26	30·58	37·
16	5·81	6·61	7·96	9·31	11·15	12·62	15·34	18·42	20·46	23·54	26·30	29·63	32·00	39·
17	6·41	7·26	8·67	10·08	12·00	13·53	16·34	19·51	21·62	24·77	27·59	31·00	33·41	40·7
18	7·02	7·91	9·39	10·86	12·86	14·44	17·34	20·60	22·76	25·99	28·87	32·35	34·80	42·3
19	7·63	8·57	10·12	11·65	13·72	15·35	18·34	21·69	23·90	27·20	30·14	33·69	36·19	43·8
20	8·26	9·24	10·85	12·44	14·58	16·27	19·34	22·78	25·04	28·41	31·41	35·02	37·57	45·3
21	8·90	9·92	11·59	13·24	15·44	17·18	20·34	23·86	26·17	29·62	32·67	36·34	38·93	46·8
22	9·54	10·60	12·34	14·04	16·31	18·10	21·34	24·94	27·30	30·81	33·92	37·66	40·29	48·2
23	10·20	11·29	13·09	14·85	17·19	19·02	22·34	26·02	28·43	32·01	35·17	38·97	41·64	49·7
24	10·86	11·99	13·85	15·66	18·06	19·94	23·34	27·10	29·55	33·20	36·42	40·27	42·98	51·
25	11·52	12·70	14·61	16·47	18·94	20·87	24·34	28·17	30·68	34·38	37·65	41·57	44·31	52·6
26	12·20	13·41	15·38	17·29	19·82	21·79	25·34	29·25	31·80	35·56	38·88	42·86	45·64	54·0
27	12·88	14·12	16·15	18·11	20·70	22·72	26·34	30·32	32·91	36·74	40·11	44·14	46·96	55·4
28	13·56	14·85	16·93	18·94	21·59	23·65	27·34	31·39	34·03	37·92	41·34	45·42	48·28	56·8
29	14·26	15·57	17·71	19·77	22·48	24·58	28·34	32·46	35·14	39·09	42·56	46·69	49·59	58·3
30	14·95	16·31	18·49	20·60	23·36	25·51	29·34	33·53	36·25	40·26	43·77	47·96	50·89	59·7

SOURCE: Abridged from Table IV of Fisher, R. A., and Yates, F., *Statistical Tables for Biological, Agricultura* *and Medical Research*, Edinburgh: Oliver & Boyd, by permission of the authors and publishers.

Table C: **Critical values of D for one-sample two-tailed Kolmogorov–Smirnov tests.**

Sample size (N)	Level of significance for D = maximum $\lvert F_0(X) - S_N(X) \rvert$				
	·20	·15	·10	·05	·01
1	·900	·925	·950	·975	·995
2	·684	·726	·776	·842	·929
3	·565	·597	·642	·708	·828
4	·494	·525	·564	·624	·733
5	·446	·474	·510	·565	·669
6	·410	·436	·470	·521	·618
7	·381	·405	·438	·486	·577
8	·358	·381	·411	·457	·543
9	·339	·360	·388	·432	·514
10	·322	·342	·368	·410	·490
11	·307	·326	·352	·391	·468
12	·295	·313	·338	·375	·450
13	·284	·302	·325	·361	·433
14	·274	·292	·314	·349	·418
15	·266	·283	·304	·338	·404
16	·258	·274	·295	·328	·392
17	·250	·266	·286	·318	·381
18	·244	·259	·278	·309	·371
19	·237	·252	·272	·301	·363
20	·231	·246	·264	·294	·356
25	·21	·22	·24	·27	·32
30	·19	·20	·22	·24	·29
35	·18	·19	·21	·23	·27
Over 35	$\dfrac{1\cdot07}{\sqrt{N}}$	$\dfrac{1\cdot14}{\sqrt{N}}$	$\dfrac{1\cdot22}{\sqrt{N}}$	$\dfrac{1\cdot36}{\sqrt{N}}$	$\dfrac{1\cdot63}{\sqrt{N}}$

SOURCE: Adapted from Massey, F. J., Jr (1951), The Kolmogorov–Smirnov test for goodness of fit, *Journal of the American Statistical Association*, **46**, 70, with the kind permission of the author and publisher.

Table D: Critical values of D for two-sample two-tailed Kolmogorov–Smirnov tests.

Level of significance	Value of D so large as to call for rejection of H_0 at the indicated level of significance, where $D = \text{maximum } |S_{n_1}(X) - S_{n_2}(X)|$
$\cdot 10$	$1 \cdot 22 \sqrt{\dfrac{n_1 + n_2}{n_1 n_2}}$
$\cdot 05$	$1 \cdot 36 \sqrt{\dfrac{n_1 + n_2}{n_1 n_2}}$
$\cdot 025$	$1 \cdot 48 \sqrt{\dfrac{n_1 + n_2}{n_1 n_2}}$
$\cdot 01$	$1 \cdot 63 \sqrt{\dfrac{n_1 + n_2}{n_1 n_2}}$
$\cdot 005$	$1 \cdot 73 \sqrt{\dfrac{n_1 + n_2}{n_1 n_2}}$
$\cdot 001$	$1 \cdot 95 \sqrt{\dfrac{n_1 + n_2}{n_1 n_2}}$

SOURCE: Adapted from Smirnov, N. (1948), Tables for estimating the goodness of fit of empirical distributions, *Annals of Mathematical Statistics*, **19**, 280–81, with the kind permission of the publisher.

Table E: Critical values of *t*

ν	Level of significance for one-tailed test					
	·10	·05	·025	·01	·005	·0005
	Level of significance for two-tailed test					
	·20	·10	·05	·02	·01	·001
1	3·078	6·314	12·706	31·821	63·657	636·619
2	1·886	2·920	4·303	6·965	9·925	31·598
3	1·638	2·353	3·182	4·541	5·841	12·941
4	1·533	2·132	2·776	3·747	4·604	8·610
5	1·476	2·015	2·571	3·365	4·032	6·859
6	1·440	1·943	2·447	3·143	3·707	5·959
7	1·415	1·895	2·365	2·998	3·499	5·405
8	1·397	1·860	2·306	2·896	3·355	5·041
9	1·383	1·833	2·262	2·821	3·250	4·781
10	1·372	1·812	2·228	2·764	3·169	4·587
11	1·363	1·796	2·201	2·718	3·106	4·437
12	1·356	1·782	2·179	2·681	3·055	4·318
13	1·350	1·771	2·160	2·650	3·012	4·221
14	1·345	1·761	2·145	2·624	2·977	4·140
15	1·341	1·753	2·131	2·602	2·947	4·073
16	1·337	1·746	2·120	2·583	2·921	4·015
17	1·333	1·740	2·110	2·567	2·898	3·965
18	1·330	1·734	2·101	2·552	2·878	3·922
19	1·328	1·729	2·093	2·539	2·861	3·883
20	1·325	1·725	2·086	2·528	2·845	3·850
21	1·323	1·721	2·080	2·518	2·831	3·819
22	1·321	1·717	2·074	2·508	2·819	3·792
23	1·319	1·714	2·069	2·500	2·807	3·767
24	1·318	1·711	2·064	2·492	2·797	3·745
25	1·316	1·708	2·060	2·485	2·787	3·725
26	1·315	1·706	2·056	2·479	2·779	3·707
27	1·314	1·703	2·052	2·473	2·771	3·690
28	1·313	1·701	2·048	2·467	2·763	3·674
29	1·311	1·699	2·045	2·462	2·756	3·659
30	1·310	1·697	2·042	2·457	2·750	3·646
40	1·303	1·684	2·021	2·423	2·704	3·551
60	1·296	1·671	2·000	2·390	2·660	3·460
120	1·289	1·658	1·980	2·358	2·617	3·373
∞	1·282	1·645	1·960	2·326	2·576	3·291

SOURCE: Abridged from Table III of Fisher, R. A., and Yates, C., *Statistical Tables for Biological, Agricultural, and Medical Research*, Edinburgh: Oliver & Boyd, by permission of the authors and publishers.

Table F: Probabilities associated with extreme values of z under the normal curve.

z	·00	·01	·02	·03	·04	·05	·06	·07	·08	·09
·0	·5000	·4960	·4920	·4880	·4840	·4801	·4761	·4721	·4681	·4641
·1	·4602	·4562	·4522	·4483	·4443	·4404	·4364	·4325	·4286	·4247
·2	·4207	·4168	·4129	·4090	·4052	·4013	·3974	·3936	·3897	·3859
·3	·3821	·3783	·3745	·3707	·3669	·3632	·3594	·3557	·3520	·3483
·4	·3446	·3409	·3372	·3336	·3300	·3264	·3228	·3192	·3156	·3121
·5	·3085	·3050	·3015	·2981	·2946	·2912	·2877	·2843	·2810	·2776
·6	·2743	·2709	·2676	·2643	·2611	·2578	·2546	·2514	·2483	·2451
·7	·2420	·2389	·2358	·2327	·2296	·2266	·2236	·2206	·2177	·2148
·8	·2119	·2090	·2061	·2033	·2005	·1977	·1949	·1922	·1894	·1867
·9	·1841	·1814	·1788	·1762	·1736	·1711	·1685	·1660	·1635	·1611
1·0	·1587	·1562	·1539	·1515	·1492	·1469	·1446	·1423	·1401	·1379
1·1	·1357	·1335	·1314	·1292	·1271	·1251	·1230	·1210	·1190	·1170
1·2	·1151	·1131	·1112	·1093	·1075	·1056	·1038	·1020	·1003	·0985
1·3	·0968	·0951	·0934	·0918	·0901	·0885	·0869	·0853	·0838	·0823
1·4	·0808	·0793	·0778	·0764	·0749	·0735	·0721	·0708	·0694	·0681
1·5	·0668	·0655	·0643	·0630	·0618	·0606	·0594	·0582	·0571	·0559
1·6	·0548	·0537	·0526	·0516	·0505	·0495	·0485	·0475	·0465	·0455
1·7	·0446	·0436	·0427	·0418	·0409	·0401	·0392	·0384	·0375	·0367
1·8	·0359	·0351	·0344	·0336	·0329	·0322	·0314	·0307	·0301	·0294
1·9	·0287	·0281	·0274	·0268	·0262	·0256	·0250	·0244	·0239	·0233
2·0	·0228	·0222	·0217	·0212	·0207	·0202	·0197	·0192	·0188	·0183
2·1	·0179	·0174	·0170	·0166	·0162	·0158	·0154	·0150	·0146	·0143
2·2	·0139	·0136	·0132	·0129	·0125	·0122	·0119	·0116	·0113	·0110
2·3	·0107	·0104	·0102	·0099	·0096	·0094	·0091	·0089	·0087	·0084
2·4	·0082	·0080	·0078	·0075	·0073	·0071	·0069	·0068	·0066	·0064
2·5	·0062	·0060	·0059	·0057	·0055	·0054	·0052	·0051	·0049	·0048
2·6	·0047	·0045	·0044	·0043	·0041	·0040	·0039	·0038	·0037	·0036
2·7	·0035	·0034	·0033	·0032	·0031	·0030	·0029	·0028	·0027	·0026
2·8	·0026	·0025	·0024	·0023	·0023	·0022	·0021	·0021	·0020	·0019
2·9	·0019	·0018	·0018	·0017	·0016	·0016	·0015	·0015	·0014	·0014
3·0	·0013	·0013	·0013	·0012	·0012	·0011	·0011	·0011	·0010	·0010

This table gives one-tailed probabilities under H_0 of z. The probabilities correspond with the shaded areas in Figures 2.2 and 4.2. Thus, for example, the z value of 1·65 has a one-tailed probability of ·05: for a two-tailed test with $\alpha = ·05$ (divided so that a probability of ·025 lies in both tails) the corresponding z value is 1·96.

Table G: Confidence limits for β_1 and β_2

	β_1			β_2		
	Upper limits	Lower limits	Lower limits		Upper limits	
N	10%	2%	1%	5%	5%	1%
25	·51	1·08	—	—	—	—
30	·44	·94	—	—	—	—
35	·39	·83	—	—	—	—
40	·35	·74	—	—	—	—
45	·31	·67	—	—	—	—
50	·29	·62	—	—	—	—
75	·20	·42	—	—	—	—
100	·15	·32	2·18	2·35	3·77	4·39
125	·12	·26	2·24	2·40	3·70	4·24
150	·10	·22	2·29	2·45	3·65	4·14
175	·09	·19	2·33	2·48	3·61	4·05
200	·08	·16	2·37	2·51	3·57	3·98
250	·06	·13	2·42	2·55	3·52	3·87

SOURCE: Modified from Pearson (1930), A further development of tests of normality, *Biometrika*, **22**, 239–48.

Table H: Critical values of F

v_1\v_2	1	2	3	4	5	6	8	12	24	∞
1	161·4	199·5	215·7	224·6	230·2	234·0	238·9	243·9	249·0	254·3
2	18·51	19·00	19·16	19·25	19·30	19·33	19·37	19·41	19·45	19·50
3	10·13	9·55	9·28	9·12	9·01	8·94	8·84	8·74	8·64	8·53
4	7·71	6·94	6·59	6·39	6·26	6·16	6·04	5·91	5·77	5·63
5	6·61	5·79	5·41	5·19	5·05	4·95	4·82	4·68	4·53	4·36
6	5·99	5·14	4·76	4·53	4·39	4·28	4·15	4·00	3·84	3·67
7	5·59	4·74	4·35	4·12	3·97	3·87	3·73	3·57	3·41	3·23
8	5·32	4·46	4·07	3·84	3·69	3·58	3·44	3·28	3·12	2·93
9	5·12	4·26	3·86	3·63	3·48	3·37	3·23	3·07	2·90	2·71
10	4·96	4·10	3·71	3·48	3·33	3·22	3·07	2·91	2·74	2·54
11	4·84	3·98	3·59	3·36	3·20	3·09	2·95	2·79	2·61	2·40
12	4·75	3·88	3·49	3·26	3·11	3·00	2·85	2·69	2·50	2·30
13	4·67	3·80	3·41	3·18	3·02	2·92	2·77	2·60	2·42	2·21
14	4·60	3·74	3·34	3·11	2·96	2·85	2·70	2·53	2·35	2·13
15	4·54	3·68	3·29	3·06	2·90	2·79	2·64	2·48	2·29	2·07
16	4·49	3·63	3·24	3·01	2·85	2·74	2·59	2·42	2·24	2·01
17	4·45	3·59	3·20	2·96	2·81	2·70	2·55	2·38	2·19	1·96
18	4·41	3·55	3·16	2·93	2·77	2·66	2·51	2·34	2·15	1·92
19	4·38	3·52	3·13	2·90	2·74	2·63	2·48	2·31	2·11	1·88
20	4·35	3·49	3·10	2·87	2·71	2·60	2·45	2·28	2·08	1·84
21	4·32	3·47	3·07	2·84	2·68	2·57	2·42	2·25	2·05	1·81
22	4·30	3·44	3·05	2·82	2·66	2·55	2·40	2·23	2·03	1·78
23	4·28	3·42	3·03	2·80	2·64	2·53	2·38	2·20	2·00	1·76
24	4·26	3·40	3·01	2·78	2·62	2·51	2·36	2·18	1·98	1·73
25	4·24	3·38	2·99	2·76	2·60	2·49	2·34	2·16	1·96	1·71
26	4·22	3·37	2·98	2·74	2·59	2·47	2·32	2·15	1·95	1·69
27	4·21	3·35	2·96	2·73	2·57	2·46	2·30	2·13	1·93	1·67
28	4·20	3·34	2·95	2·71	2·56	2·44	2·29	2·12	1·91	1·65
29	4·18	3·33	2·93	2·70	2·54	2·43	2·28	2·10	1·90	1·64
30	4·17	3·32	2·92	2·69	2·53	2·42	2·27	2·09	1·89	1·62
40	4·08	3·23	2·84	2·61	2·45	2·34	2·18	2·00	1·79	1·51
60	4·00	3·15	2·76	2·52	2·37	2·25	2·10	1·92	1·70	1·39
120	3·92	3·07	2·68	2·45	2·29	2·17	2·02	1·83	1·61	1·25
∞	3·84	2·99	2·60	2·37	2·21	2·09	1·94	1·75	1·52	1·00

SOURCE: Abridged from Table V of Fisher, R. A., and Yates, F., *Statistical Tables for Biological, Agricultural, and Medical Research*, Edinburgh: Oliver & Boyd, by permission of the authors and publishers.

Critical values of F (continued)

v_2 \ v_1	1	2	3	4	5	6	8	12	24	∞
1	4052	4999	5403	5625	5764	5859	5981	6106	6234	6366
2	98·49	99·01	99·17	99·25	99·30	99·33	99·36	99·42	99·46	99·50
3	34·12	30·81	29·46	28·71	28·24	27·91	27·49	27·05	26·60	26·12
4	21·20	18·00	16·69	15·98	15·52	15·21	14·80	14·37	13·93	13·46
5	16·26	13·27	12·06	11·39	10·97	10·67	10·27	9·89	9·47	9·02
6	13·74	10·92	9·78	9·15	8·75	8·47	8·10	7·72	7·31	6·88
7	12·25	9·55	8·45	7·85	7·46	7·19	6·84	6·47	6·07	5·65
8	11·26	8·65	7·59	7·01	6·63	6·37	6·03	5·67	5·28	4·86
9	10·56	8·02	6·99	6·42	6·06	5·80	5·47	5·11	4·73	4·31
10	10·04	7·56	6·55	5·99	5·64	5·39	5·06	4·71	4·33	3·91
11	9·65	7·20	6·22	5·67	5·32	5·07	4·74	4·40	4·02	3·60
12	9·33	6·93	5·95	5·41	5·06	4·82	4·50	4·16	3·78	3·36
13	9·07	6·70	5·74	5·20	4·86	4·62	4·30	3·96	3·59	3·16
14	8·86	6·51	5·56	5·03	4·69	4·46	4·14	3·80	3·43	3·00
15	8·68	6·36	5·42	4·89	4·56	4·32	4·00	3·67	3·29	2·87
16	8·53	6·23	5·29	4·77	4·44	4·20	3·89	3·55	3·18	2·75
17	8·40	6·11	5·18	4·67	4·34	4·10	3·79	3·45	3·08	2·65
18	8·28	6·01	5·09	4·58	4·25	4·01	3·71	3·37	3·00	2·57
19	8·18	5·93	5·01	4·50	4·17	3·94	3·63	3·30	2·92	2·49
20	8·10	5·85	4·94	4·43	4·10	3·87	3·56	3·23	2·86	2·42
21	8·02	5·78	4·87	4·37	4·04	3·81	3·51	3·17	2·80	2·36
22	7·94	5·72	4·82	4·31	3·99	3·76	3·45	3·12	2·75	2·31
23	7·88	5·66	4·76	4·26	3·94	3·71	3·41	3·07	2·70	2·26
24	7·82	5·61	4·72	4·22	3·90	3·67	3·36	3·03	2·66	2·21
25	7·77	5·57	4·68	4·18	3·86	3·63	3·32	2·99	2·62	2·17
26	7·72	5·53	4·64	4·14	3·82	3·59	3·29	2·96	2·58	2·13
27	7·68	5·49	4·60	4·11	3·78	3·56	3·26	2·93	2·55	2·10
28	7·64	5·45	4·57	4·07	3·75	3·53	3·23	2·90	2·52	2·06
29	7·60	5·42	4·54	4·04	3·73	3·50	3·20	2·87	2·49	2·03
30	7·56	5·39	4·51	4·02	3·70	3·47	3·17	2·84	2·47	2·01
40	7·31	5·18	4·31	3·83	3·51	3·29	2·99	2·66	2·29	1·80
60	7·08	4·98	4·13	3·65	3·34	3·12	2·82	2·50	2·12	1·60
120	6·85	4·79	3·95	3·48	3·17	2·96	2·66	2·34	1·95	1·38
∞	6·64	4·60	3·78	3·32	3·02	2·80	2·51	2·18	1·79	1·00

TABLE I: Critical values of H for the Rayleigh test

N	$\alpha = \cdot05$	$\alpha = \cdot01$
6	2·857	4·080
7	2·882	4·156
8	2·901	4·201
9	2·910	4·250
10	2·919	4·290
11	2·926	4·320
12	2·932	4·344
13	2·937	4·365
14	2·941	4·383
15	2·945	4·398
16	2·948	4·412
17	2·951	4·423
18	2·954	4·434
19	2·956	4·443
20	2·958	4·451
21	2·960	4·459
22	2·961	4·466
23	2·963	4·472
24	2·964	4·478
∞	2·996	4·605

Appendix 1: Basic mathematical operations and statistical symbols

1. Operators

The four most basic mathematical operations should be familiar. These are:

+ meaning add $\div$ or $/$ meaning divide.
− meaning subtract   and x or $\cdot$ meaning multiply.

In algebra, the multiplication sign is usually omitted so that $A \times B = C$ is written $A\,B = C$.

Equally familiar will be four other operations,

X^N meaning the Nth power of X

$\sqrt[N]{X}$ meaning the Nth root of X

$\log X$ meaning the common logarithm (to base 10) of X

or $\ln X$ meaning the natural logarithm (to base e) of X

and N^X meaning N raised to the exponent X.

Powering provides a particularly useful notation because this operation may also be used to signify a root and a reciprocal. For example

$$X^{\frac{1}{3}} = \sqrt[3]{X}$$

$$X^{-2} = \frac{1}{X^2}$$

$$X^{-\frac{1}{2}} = \frac{1}{\sqrt{X}}$$

Note that $X^0 = 1$ (by definition).

Another operator which is frequently used by statisticians is the factorial sign !; it has the following meaning:

$$N! = (N)\,(N-1)\,(N-2)\,(N-3)\ldots(2)\,(1)$$

Thus if $N = 5$, then

$$N! = 5 \times 4 \times 3 \times 2 \times 1 = 120$$

Note that $0! = 1$ (by definition). We also have ∞ = infinity and $|X|$ = the absolute value of X: thus $|-4| = 4$.

N.B. When making notes, it is important to distinguish the upper and lower case form of the letter X from the operation $\times$ (to multiply).

2.   *Subscripts*

In many instances one has a series of observations on a particular variable, such as the temperature at a weather station read hourly over a twenty-four hour period. Rather than going through the laborious procedure of attaching a label to each observation, we may give all the observations the same label, identifying them with different subscripts. Thus we may replace four identifiers A, B, C, D with a one-dimensional array X_i ($i = 1, 2, 3$ and 4). Subscripts are a particularly valuable shorthand. Equally valid are superscripts where the identification is made above the variable, but these are less frequently used because they can be confused with the notation for powering.

More than one subscript may be attached to a variable, and in such fields as interregional input-output analysis, several subscripts may be used. Here we need only concern ourselves with two dimensional arrays (commonly known as matrices) which are identified by two subscripts. The rule is that the first subscript identifies the row, the second subscript identifies the column. Thus

$$[X] = X_{ij} = \begin{bmatrix} X_{11} & X_{12} & X_{13} & \cdots & X_{1N} \\ X_{21} & X_{22} & X_{23} & & \cdot \\ X_{31} & X_{32} & X_{33} & & \cdot \\ \cdot & & & & \cdot \\ \cdot & & & & \cdot \\ \cdot & & & & \cdot \\ X_{M1} & \cdots & \cdots & \cdots & X_{MN} \end{bmatrix}$$

By convention the letters i and j (and sometimes k and ℓ) are used as subscripts, whilst M and N are frequently used to signify the number of elements in a column and row respectively.

3.   *Summation and product signs*

Statisticians frequently make use of Greek letters to denote particular coefficients, measures and so on. Appendix 2 provides a summary of the Greek alphabet, together with the pronunciation and statistical usage of certain letters. One Greek letter which is of particular importance to statisticians is upper-case sigma (written Σ). This symbol means 'the sum of' and refers to the subscript written below the sigma sign. For example, taking the four element array used in the previous section

$$\sum_{i=1}^{4} X_i = X_1 + X_2 + X_3 + X_4$$

The numbers below and above the sigma sign indicate the first and last values (inclusive) in the series that is being summed. Where one is summing only part of an array, as in the case

$$\sum_{i=2}^{4} X_i = X_2 + X_3 + X_4$$

it is essential to indicate the first and last values, but where one is summing the whole of a defined series, it is normal to omit these values, so that one has

$$\sum_{i} X_i \text{ or even more briefly } \Sigma X_i$$

The summation sign may also be used with matrices to indicate the summing of rows, columns or both. Take the case of a simple 3 x 3 matrix:

$$X_{ij} = \begin{bmatrix} 3 & -2 & 4 \\ 1 & 9 & -3 \\ 7 & 2 & -4 \end{bmatrix}$$

If one wishes to sum across the rows, then for the first row this involves adding X_{11} to X_{12} and X_{13}. Note that the first subscript remains constant at one while the second subscript (j) varies from one to three as one jumps from column to column, hence we wish to place the second subscript against the summation sign to sum across a row. The same applies to the second and third row. Here

$$\sum_{j=1}^{3} X_{ij} = \begin{bmatrix} 5 \\ 7 \\ 5 \end{bmatrix}$$

Similarly we sum for the *i*s to obtain the column totals so that

$$\sum_{i=1}^{3} X_{ij} = [11 \quad 9 \quad -3]$$

Summing the whole matrix,

$$\sum_{i=1}^{3} \sum_{j=1}^{3} X_{ij} = 17$$

The product sign, Π, is used in a manner identical to the summation sign, except that the elements are multiplied instead of being added together. Hence

$$\prod_{i=1}^{4} = X_1 \times X_2 \times X_3 \times X_4$$

Appendix 2: Statistical notation and the Greek alphabet

Lower case (LC)	Upper case (UC)	Name	Corresponding English sound or letter	Common statistical usage
α	A	alpha	a	(LC) Level of significance: Probability of a Type I error.
β	B	beta	b	(UC) An important statistical distribution: (LC) Probability of a Type II error: also β_1 is a measure of skewness; and β_2 is a measure of kurtosis.
γ	Γ	gamma	g	(UC) An important statistical distribution.
δ	Δ	delta	d	(UC) Change in parameter over a given time period.
ϵ	E	epsilon	e (short)	(LC) A random error variate.
λ	Λ	lambda	l	(LC) The mean and variance of the Poisson distribution.
μ	M	mu	m	(LC) The mean of a population.
ν	N	nu	n	(LC) The number of degrees of freedom.
π	Π	pi	p	(LC) 2·173 . . . of spherical geometry; also any moment of a distribution: (UC) The product of . . .

Lower case (LC)	Upper case (UC)	Name	Corresponding English sound or letter	Common statistical usage	
ρ	P	rho	r	(LC)	Population correlation coefficient.
♀		koppa	q	(LC)	The population coefficient of medial correlation.
σ	Σ	sigma	s	(LC)	population standard deviation:
				(UC)	The summation of . . .
τ	T	tau	t	(LC)	Kendall's rank correlation coefficient.
ϕ	Φ	phi	ph	(LC)	A coefficient of association.
χ	X	chi	k	(LC)	χ^2 is a commonly used statistical distribution and the name of a test based on this distribution.

Index